CHONX STIX

Magnetic modelling blocks for crystals, molecules and other structures

ISBN-13: 978-1514390290

ISBN-10: 1514390299

All pictures are from www.wikipedia.com and www.google.com

INDEX

The atoms
Carbon, Hydrogen, Oxygen, Nitrogen, Sodium, Chlorine

Molecules in the air
Nitrogen, (N2)
Oxygen (O2)
Ozone (O3)
Water (H2O)
Carbon dioxide (CO2)
Carbon monoxide (CO)
Nitrous oxide (N2O)
Sulfur dioxide (SO2)
Nitric oxide (NO
Nitrogen dioxide (NO2)

Molecules of fuel

Methane (CH4)
Ethane (CH_3-CH_3)
Propane (C3H3)
Butane ($CH_3CH_2CH_2CH_3$)
Pentane (C_5H_{12})
Hexane ($CH_3(CH_2)_4CH_3$)
Heptane ($H_3C(CH_2)_5CH_3$ or C_7H_{16})
Octane (C8H18)
Hydrogen peroxide

Sugar molecules
Glucose ($C_6H_{12}O_6$)
Fructose ($C_6H_{12}O_6$)
Sucrose ($C_{12}H_{22}O_{11}$)
Cellulose (C6H10O5)n

Alcohols
Methanol, also known as wood alcohol (CH_3OH)
Ethanol also called ethyl alcohol, or grain alcohol, (CH3CH2OH
Glycerol

Ethers
Dimethyl ether

Aldehydes
Formaldehyde

Ketones
Acetone

Acids, bases and salts
Formic acid (HCOOH)
Acetic acid (CH_3COOH)
Vitamin C or L-ascorbic acid, or simply ascorbate
Sodium hydroxide, also known as caustic soda, or lye (NaOH).
Hydrogen Chloride (HCl)
Sodium Chloride (NaCl)

Esters
Ethyl acetate

Fat Molecules
Saturated,   Unsaturated,   Trans
Cholesterol,
Omega-3 fatty acids
Tocopherol, or Vitamin E,

Protein molecules
Alanine
Solvent molecules
Benzene (C_6H_6)
Cyclohexane (C_6H_{12})
Toluene
Acetone ($CH_3(CO)CH_3$)
Turpentine
Tetrachloroethylene ($Cl_2C=CCl_2$)

Molecules of color
Methyl red
Methyl orange
Methyl yellow,
Methyl violet
Chlorophyll
Carotene

Molecules of smell and taste
Menthol
Geraniol
Linalool
Diacetyl
Decanoic acid, or capric acid
Ethyl acetate
Butanone,
Butyric acid
n-Butyl acetate,
Amyl butyrate
Diallyl disulfide
Asparagusic acid

Molecules for emotions
Dopamine ($C_6H_3(OH)_2$-CH_2-CH_2-NH_2)
ATP ($C_{10}H_8N_4O_2NH_2(OH)_2(PO_3H)_3H$)
Epinephrine or adrenaline
Testosterone,
Estrogens
Hemoglobin,
DDT ($ClC_6H_4)_2CH(CCl_3$)

Cyanides
Hydrogen cyanide (HCN)
Prussian blue ($Fe_7(CN)$)
Sodium cyanide (NaCN)

Explosive molecules
Nitro compounds
Nitrate
Nitroglycerin (NG),
Trinitrotoluene (TNT), ($C_6H_2(NO_2)_3CH_3$)

Drug molecules

Stimulants
Cocain
Methamphetamine ($C_{10}H_{15}N$)
Caffeine
Nicotine
Amphetamine
Methylphenidate (MPH)

Analgesics
Aspirin
Paracetamol or acetaminophen (USAN),
Codeine or methylmorphine (INN) ($C_{18}H_{21}NO_3$)
Morphine $\underline{C}_{17}\underline{H}_{19}\underline{NO}_3$

Antibiotics
Penicillin (abbreviated PCN) (R-$C_9H_{11}N_2O_4S$), where R is a variable side chain.
Tetracycline ($C_{22}H_{24}N_2O_8$)

Sedatives
Barbiturates
Anti-insomnia
Ambien or Zolpidem

Anesthetics
Procaine (Nonocaine)

Anti-impotence agents
Sildenafil citrate, sold under the names Viagra, Revatio

Hallucinogens
Tetrahydrocannabinol, (THC)
Lysergic acid diethylamide,(LSD),

Crystals and Molecules in the earth

Diamond
Graphite
Halite

Molecules in the laboratory

Styrene ($C_6H_5CH=CH_2$)
Plexiglass also called *Methylmethacrylate* ($CH_2=C(CH_3)COOCH_3$)
Nylon
Polytetrafluoroethylene (PTFE) better known as Teflon
Tetraoxygen also called oxozone (O_4)
Red oxygen (O4) in a crystal lattice.
Tetrahedrane (C_4H_4)
Cubane (C_8H_8)
Dodecahedrane ($C_{20}H_{20}$)
Basketane ($C_{10}H_{12}$)
Fullerene
Graphene

CHONX STIX are magnetic pieces that represent the atoms C,H,O,N as well as metal and halogen atoms. Modelling crystals and molecules with CHONX STIX offer many advantages over modelling atoms using the traditional basll and stick approach.

In addition to modelling crystals and molecules, CHONX STIX can be used to model plants, animals, buildings and interesting structures limited only by imagination. CHONX STIX are magnetic. The magnets align themselves so that very complex structures like the diamond crystal can be modelled by a child. Just gently pick up the pieces that are connected magnetically and gently shake it until the pieces spring into alignment.

Crystals and Molecules

The ingredients for crystals and molecules are the atoms that are cooked in stars.

A **crystal** or crystalline solid is a solid material whose constituent atoms, molecules, or ions are arranged in an ordered pattern extending in all three spatial dimensions. In addition to their microscopic structure, large crystals are usually identifiable by their macroscopic geometrical shape, consisting of flat faces with specific, characteristic orientations.

Examples of large crystals include snowflakes, diamonds, and table salt. Most inorganic solids are not crystals but **polycrystals**, i.e. many microscopic crystals fused together into a single solid. Examples of polycrystals include most metals, rocks, ceramics, and ice. A third category of solids is **amorphous solids**, where the atoms have no periodic structure whatsoever. Examples of amorphous solids include glass, wax, and many plastics.

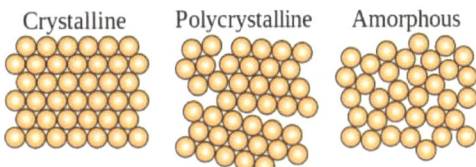

Crystalline   Polycrystalline   Amorphous

A **molecule** is an electrically neutral group of two or more atoms held together by chemical bonds. Molecules are distinguished from ions by their lack of electrical charge.

Organic molecules arc based on carbon atoms that are dressed in Hydrogen and decorated with mainly Oxygen and Nitrogen atoms. They are made by Life, either in living bodies, or by scientists in laboratories. Althogh many of the molecules have very similar shapes, they have very different functions. Like different types of animals, many have distinctive head and tails with a body in the middle that varies in size. The organic molecules are like keys that all look the same, but all open differnt locks.

The atoms

Atoms are composed of positively charged protons an equal amount of negativelly charged electrons. Each proton is bonded to one electron, cancelling their charges and making the atom neutral. Electrons of atoms can be paired or unpaired. Electrons from one atom bond with electrons from another atom to form molecules. When electrons of one atom bond with electrons from other atoms, they form *single bonds*. When atoms are bonded by 2 single bonds, the bond is called a *double bond*. When atoms are bonded by 3 single bonds, the bond is called a *triple bond*. Think of bonds between 2 atoms like two acrobats holding each other using their hands and feet. Different atoms are grouped together in families.

The family of atoms led by hydrogen have 1 or more unpaired electrons that are so loosely held that it easily breaks away from its proton leaving the atoms as positive ions in a sea of free electrons. This gives these atoms *metallic* properties of luster, malleability and thermal and electrical conductivity.

The family of atoms led by *carbon* have 2 unpaired electrons, making the atoms ideal for forming bonds with atoms on either side resulting in strings of atoms called polymers seen in plants and plastics. When carbon forms bonds with other carbons, the bonds are strong enough to form stable molecules but weak enough to be easily broken by external forces like temperature. This allows types of molecules to form that are necessary for life

The family of atoms led by *nitrogen* have 3 unpaired electrons allowing them to bond in such a way as to link the polymer carbon strings into 3 dimensional structures like proteins. The 3 bonds of the atoms form bonds so tight that they are explosive when broken.

The family of atoms led by *oxygen* have 2 unpaired electrons that cut and tear atoms from their molecules causing burning, fire and rust, or form bonds with atoms forming molecules such as water and alcohol. The bonds formed by oxygen are strong enough not to be easily broken by external forces like temperature.

The family of atoms called Halogens are led by *fluorine* and have 1 unpaired electron that is so tightly held that when they pair with other unpaired electrons of other atoms, they tear the atoms apart, stealing their electron. When this happens, these atoms become negative charged ions with electrical properties. Atoms of this family, when ionized, bond easily with positively charged metal ions that have lost their electrons forming salts.

The family of atoms called Noble gases are led by *helium*.and have 0 unpaired electrons. Their paired electrons are very tightly held and do not bond with any other atoms.

All atoms of the same family have the same number of unpaired electrons and have similar shapes and properties. In the figure below, unpaired electrons are shown elongated, and paired electrons are shown spherical.

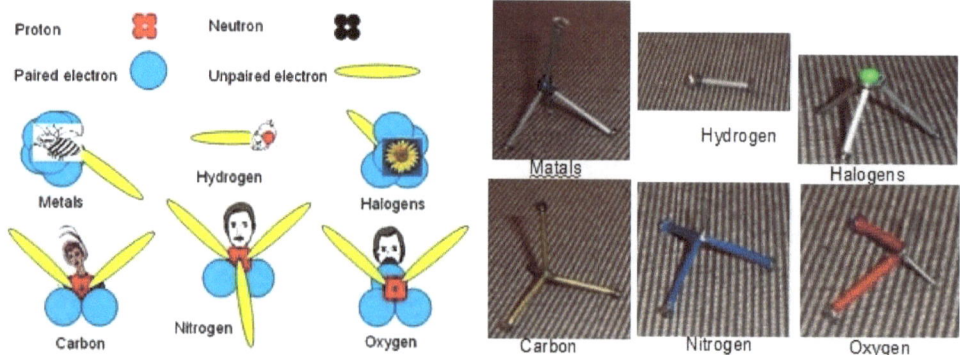

Hydrogen (H) is the most abundant element forming almost 75% of the universe. Hydrogen is the smallest atom with 1 proton and 1 unpaired electron. In the sun, hydrogen atoms are fused together to form all the other atoms. Hydrogen forms bonds with its one and only unpaired electron with most other types of atoms to form many types of molecules. Molecules with hydrogen form a very gently bond between them called the hydrogen bond. It is what holds water together in a drop. Because all other atoms are made from hydrogen in the sun, it is like a baby and has a **baby**'s face.

Hydrogen can easily break apart into its component proton and electron when tightly held by other atoms in the molecule and when sufficiently bombarded by other molecules. Hydrogen losing its protruding proton is the cause of acid reactions, like a sour baby losing his pacifier. The heat and light from stars arise mainly from the fusion of hydrogen into helium. Hydrogen forms more compounds than any other element even though it can bond with only one other element and only with a single bond.

Metals

Metals have one or more very loosely held unpaired electrons. They are loosely held by the atom because they are crowded out by the many paired electrons that metal atoms have. When electrons break away, they leave a positive charged metal ion swimming in a negatively charged sea of electrons. Electric current flows when the sea of electrons all move in the same direction like a swarm of bees. Metals atoms are like **bees** that too easily lose their stinger.

Carbon Family

Elements in this family include **Silicon (Si)**, **Germanium (Ge)**, **Tin (Sn)**, and **Lead (Pb)**. They all have 2 tightly held unpaired electrons which form chains with other molecules to make string like compounds that loop and form sheets and crystals. This is the shape that allows the closest packing of spheres allowing the hardest crystal, the diamond, to form. Their 2 unpaired electrons, like hands, form gentle bonds with most other types of atoms. Carbon forms chains with other carbon atoms that are the backbone of all life. Hydrogen is very attracted to carbon, because carbon is like a woman, forming gentle bonds and attracting hydrogen. It has a **woman**'s face.

Carbon (C) is the 3rd most available element in the universe and is found in its free form in nature as *diamonds, graphite* and *coal*. The shape of the atom is a tetrahedron with 4 corners. There are **2 unpaired electrons** to form 2 bonds with other atoms in 2 directions forming chain structures. In addition there are 2 paired electrons that form bonds with other atoms that dress and cover the carbon chains. It is these carbon chains that make up and feed all of life.

Sheets of carbon atoms form graphite. Like randomly aligned snowflakes, graphite forms a slippery material similar to snow and ice. The graphite sheets are indented much like corrugated cardboard and when enough pressure is applied to align them and lock them to each other, they form 3 dimensional structures as displayed by diamond crystals.

Carbon forms compounds with hydrogen called **hydrocarbons** (petroleum) that fuel our civilization. C is an essential component for life as it easily combines with itself and with many other elements allowing the formation of rings and long chain molecules required for complex development of life. The figures below show single and double bonds between the carbon atoms. As the carbon chain gets longer, the molecules change from gases to liquids and then to solids. Earth *gases* like *methane* have chains with up to 5 carbons. Liquid *gasoline* has up to 8 carbons, *fuel oils* like *kerosene* and *diesel* have over 8 carbons, *lubricating* oils and *grease* have over 20 and solids like *paraffin wax* have 20-40 carbon atoms in their chain. *Rubbers* and *plastics* are polymers made up of an indefinite large number of carbons in a chain.

Just as carbon is required for the development of life, **Silicon (Si)**, its heavier brother, is required for the development of computers. Si forms 70% of the land mass of earth in the form of *sand and rocks and glass, asbestos, mica, clay, talc, quartz, topaz, garnet,* and *agate*. Silicon in the form of SiO_2, the oxide of silicon, unlike CO_2, is a crystalline solid because of the single bonds it forms with O (-O-Si-O-Si-O-) forming a network like diamond. It creates either massive hard and brittle crystals or powders consisting of tiny crystals too small to be seen with an optical microscope. They form *boulders* a few meters in diameter break up into smaller rock pieces called *gravel*. *Sand* cemented to form *sandstone* by nature and *concrete* by man. *Mud* and even smaller pieces called *silt* ground by glaciers form *mudstones* and *siltstones*. *Clay* pieces form *claystones* by nature and *porcelains* by man. *Granite*, solidified molten Si crystals hardened by the pressures of the earth break down into rock which breaks down into sand. *Limestone,* sediment of sand cemented with calcium carbonate ($CaCO_3$) from *seashells* forms. *Marble* is recrystalized limestone buried deep under ground under the oceans.

Silicones are non toxic inorganic polymer chain of SiOSiOSiO... By adjusting the size of the chain, fluids, resins, and rubbers used in lubricants, water repellents, waxes and polishes and non static coatings are produced. They are far more resistant to oxidation than organic polymers because the Si-O bond is stronger than the C-C bond. The chain is easily twisted and rotates preventing close contact. This causes a lower freezing point, useful for motor oils. It is used as silly putty, bathtub caulk, and breast implants.

Silicate fibers are similar to cellulose and cellophane which are based on chains of carbon.

Nature made rock fibers like *mica* (Metals-Si8-O20-(OH)4) and *asbestos* (Metals-Si8-O22-(OH)2) are based on chains of silicon and have metals such as Ca, Na, Mg, Fe and Al on their chains, making them toxic when inhaled.

Man-made rock fibers like *rock wool* made like cotton candy, *fiberglass* made like cloth and *fiber cement* made like cardboard do not have metals on their chains.

Silica Gel is formed when sodium carbonate from sea shells and silicon dioxide from sand melt and the carbons and silicon atoms trade positions. They form sodium silicate and carbon dioxide: ($Na_2CO_3 + SiO_2 \rightarrow Na_2SiO_3 + CO_2$) While $CaCO_3$ dissolves only slightly in water making what is called *hard water*, $NaSiO_3$ dissolves readily in water forming a basic solution called *water glass*. When water glass is heated to 100–105 °C, the water evaporates leaving behind a residue of granular glass called Silica Gel with pores 2.4 nanometers diameter and with a very strong affinity for water molecules. Silica Gel granules are used as a desiccant to keep things dry.

Nitrogen Family

Elements in this family (**N, P, As, Sb** and **Bi**). have a shape with 3 tightly held unpaired electrons which form bonds with other molecules.

Nitrogen (N) is the 4th most available element in the human body and makes up 80% of air. It is used in *amines* which form *amino acids*, the building blocks of proteins, and *nucleic acids*, the building blocks of *DNA*. These atoms offer 3 unpaired electrons which bond with other atoms in 3 directions and tie 2 dimensional carbon chains into complex 3 dimensional forms called *proteins*. Nitrogen forms very tight triple bonds that when broken release great amounts of explosive energy as in explosives and bombs.

Like a net that spiders build to trap and capture insects, or like the net society builds to capture and hold information, nature builds net like structures using nitrogen to capture and store light energy to be used by plants and to hold and carry oxygen so that animals can burn the plants for energy.

Chlorophyll in plants like a kite is a ring of 4 nitrogen atoms in a web of carbon with a long trailing chain. 4 nitrogen atoms hold a *magnesium* atom which on *absorbing photons* causes electrons to hop from one atom to another down the tail. This flow of electrons is used as an energy source much like a current of electricity from a battery or from lightning hitting a kite.

Haemoglobin in animals is a similar structure to chlorophyll. The 4 nitrogen atoms hold an *iron* atom which *carries an oxygen* atom to cells which remove the oxygen and use it to burn the carbohydrates for energy that the plants produced from light.

Proteins are chains of amino acids. Amino acids are molecules with two heads. A COOH acid head and a NH2 head on a tail or chain made up of C, H, O and N atoms. All proteins in the human body are made up of only 20 amino acids. 11 of the 20 can be produced by our DNA. 9 must be produced by DNA of other life forms, making us dependant on them. All enzymes, hormones, and tissues in the human body, except fat, are made of protein.

Oxygen Family

Elements (**O, S, Se, Te,** and **Po**) have a shape with 2 tightly held unpaired electrons which like hands form bonds with other molecules. At the same time the 2 unpaired electrons are so tightly held by the nucleus, they often break apart molecules like a karate fighter.

Oxygen (O) is the 5th most abundant element in the universe. It makes up 20% of air and is essential to sustain and fuel life. Oxygen offers 2 unpaired electrons for forming 2 very tight bonds with many types of atoms. They form oxides with most metals (except Gold, Platinum and Mercury) corroding the metals forming *rust* and *tarnish*. It bonds with hydrogen to make water. Because oxygen is like a fireman or like a policeman, it has a **man**'s face.

2 atoms join to form molecules of O_2. 3 atoms form *ozone* (O_3) that like umbrellas shield us from harmful radiation. Oxygen bonds with 2 hydrogen atoms forming water (H_2O) vital for life. It bonds with carbon forming carbon dioxide (CO_2) to make dry ice, to feed the plants and to bubble our soda. Too much CO_2 in the air causes the air to act like glass (SiO_2) in a greenhouse to trap the heat causing global warming.

It bonds with nitrogen forming laughing gas (N_2O) to make us light headed. The centre N alternates from having 2 double bonds to having a single and a triple bond.

It forms nitric oxide (NO) used by mammals as a cell signalling molecule. It forms *nitro-glycerine*, a powerful explosive.

The shape of oxygen with 2 unpaired electrons, like hands, allows oxygen to easily bond on each side. The atom's shape also causes oxidation of molecules by cutting up and tearing them apart when it is freely rotating and swinging its hands.

Oxygen atoms cut up long chains of carbon built by plants into their smaller constituents. Like the cracking sound emitted when breaking twigs, energy is released when breaking carbon chains. This energy is used by animals when they break down carbon chains from the plants they eat and digest. The end products of this burning is CO_2 and H_2O which are released into the air like smoke from a fire. The plants use CO_2, H_2O and sunlight to rebuild the long carbon chains so that the production of energy and materials necessary for animal life can be sustained by the cycle.

When the energetic bonding capability of oxygen is regulated and controlled, complex stable molecule chains form, like carbohydrates and acids.

When hydrocarbon chains are capped with OH, **carbohydrates**, the building blocks of plants and the fuel of animals like *alcohols* and *sugars* are formed. Animals use oxygen from air to break down and burn the carbon chains of carbohydrates from plants into H_2O and CO_2 when they eat and digest the plants. The plants, using sunlight energy, take H_2O and CO_2 from air to rebuild these long chains of carbohydrates and return the oxygen to the air.

CO_2 is the by-products of machines when they burn hydrocarbons and a by-product of animals when they burn carbohydrates. One car emits about 2000kg of CO_2 a year while a human, about 360kg a year. Looking at it another way, a human emits per day about 900g of CO_2 but a hamburger has a CO_2 footprint of about 3000 g and a litre carbonated drink like Coca-Cola has 4.4 g of CO_2. Too much CO_2 in the air causes a global warming greenhouse effect.

CO_2, like SiO_2 is a bit strange in its properties. It has a boiling / condensing point lower than the freezing / melting point causing the solid form (dry ice) to boil into gas before melting into liquid. Like glass, it acts as a greenhouse causing global warming. When there is too much CO_2 in the air, nature eventually goes into deep freeze with an ice age and puts an end to any more CO_2 production for a while. It`s nature's way of keeping life and machines in check. CO_2 is a gas at room temperatures because oxygen forms double bonds with carbon (O=C=O) creating small single molecules.

Alcohols are hydrocarbon tails (R) with (OH) heads. Chained alcohols form *sugars*, and chained sugars form *starches*. When the tail is very long, containing more than 10,000 carbon atoms, *cellulose* is formed. Cellulose, also known as dietary fibre, is the structural component of green plants forming 30% of all plant matter. Wood has 50% cellulose while cotton has 90%. Cellulose is used to make *paper* sheets, *cellophane* sheets and *rayon* strings.

When sugars are oxidized in a controlled manner, they break down into alcohols. Sugars are broken down into alcohols by bacteria when they *ferment* fruit juices. When alcohol is further oxidized in a controlled manner, the hydrocarbon chains (R) are capped with (COOH) heads forming organic *acids* the building blocks of fats and proteins.

The strong bonds of oxygen give acids and bases their functionality.

When hydrogen is held on a molecule by 2 oxygen atoms, as it is in **acids** (COOH), the bond to hydrogen is so strong that when hydrogen is knocked away, the 2 oxygen atoms keep and hold hydrogen`s electron behind. The protruding proton breaks off breaking hydrogen in two and leaving the molecule as a negatively charged ion.

When hydrogen is held on a metal molecule by an oxygen, as it is in **bases** (OH), the bond to the molecule is such that when the oxygen and hydrogen are torn away, the tightly bonding oxygen tears out an electron from the metal leaving the metal a positively charged ion. This allows ionic and electrolytic reactions of acids and bases to take place and is necessary for life.

Solid *fats* and liquid *oils* are composed of 3 chains of carbon connected in the middle by a 4th shorter carbon chain of 3 carbons as shown in the figure above. Acids combine with the alcohol *glycerol* to form *triglycerides*, ester molecules called **fats and oils**. These fat molecules are essential for forming cell walls and membranes within the cells, as well as for insulating and cushioning it from outside threats. Animals use reserve fat as fuel that the cells burn for energy.

Acids and alcohols combine to form fragrant and tasty **ester** compounds (RCOOR') as shown in the figure above. Depending on the length of the acid's tail, and the length of the alcohol's tail, molecules resembling strings of different lengths are formed. Where the acid and alcohol join, an oxygen atom pinches the string with a double bond. Just like different musical notes are produced from a vibrating string depending on its length and where it is pinched, different smells and tastes are produced by these ester molecules.

Ester strings are formed that can be easily joined together by chemists to form polymers much like plants grow cellulose. These polymers called **polyesters** can be woven and spun into threads and fabrics much like cellulose in cotton is. Esters can be designed for specific qualities by choosing the right alcohol and acid. Many interesting materials can be formed such as very strong *Mylar* sheets that make bullet proof vests.

When the alcohol head is on a metal atom, esters called **soaps** are produced. The metal end of the ester string easily dissolves in water. The hydrocarbon tail from the acid easily dissolves in fat and oil. The strings surround oil droplets and like anchors or brooms allow the metal end dissolved in water to drag and flush the oil droplet away.

When 2 hydrocarbon tails are linked by an oxygen atom (ROR'), colourless flammable liquids used as solvents and anaesthetics called **ethers** are formed. They are pleasant smelling resembling alcohols and occur naturally in starches and sugars. They are widely used in industry and in making pharmaceuticals.

When 2 hydrocarbon tails are linked by 2 oxygen atoms (ROOR'), **peroxides** used in the chemistry industry are formed. When the tails consist of only 1 hydrogen, *hydrogen peroxide* (HOOH) used as a bleach and disinfectant is formed. When hydrogen peroxide comes into contact with blood protein, it bubbles into water (H_2O) and oxygen gas (O_2).

When certain alcohols are oxidized, highly reactive compounds called **aldehydes** are formed with double bonds to the oxygen atom. They are used in manufacturing *resins*, *dyes*, and *organic acids*. *Formaldehyde* from methyl alcohol is a colourless toxic water soluble gas used as a disinfectant and preservative, and in the manufacture of resins and plastics.

When CO atoms link 2 hydrocarbon tails (R'COR), compounds called **ketones** like *acetone* are produced. Acetone is a colourless, volatile, highly flammable liquid that is widely used as a solvent, paint thinner, and nail-polish remover.

When the acids contain nitrogen, nitrogen's properties come out. These acids called **amino acids** form chains called **proteins**. The much smaller nucleic acids form twisting chains as seen in DNA. These amino acids are chains similar to esters. Like esters, they are formed by nature into materials like *wool* and *silk*, and by chemists into *poly amides* like *nylons*.

Sulphur (S), oxygen's heavier brother, is used for making bonds with its 2 unpaired electrons. Just like Oxygen bonds atoms together to form stable molecules, Sulphur bonds molecules together to form stable comounds. Sulphur is used in the *vulcanization of rubber* by cross linking the individual polymer chains. **Selenium (Se)** conducts electricity better in the light than in the dark so it is found in *photocells*, electrical components that detect light.

Halogens

Elements in this family have a shape with one very tightly held unpaired electron which forms very strong bonds with other molecules. They are the gases **Fluorine (F)** and **Chlorine (Cl)**, the liquid **Bromine (Br)** and the solid **Iodine (I)**. They are more reactive than the alkali metals and even light will cause a reaction. They react with metals forming salts. Because of their attracting powers, they are like attractive **flowers**.

Their shape allows the lone unpaired electron of the halogen atom to have an empty space the size of an unpaired electron right beside it. When an unpaired electron from another atom (like hydrogen) happens to be bumping into the halogen, the 2 unpaired electrons from each of the two atoms synchronize and pair together like a snap button, making molecules like HCl. HCl has the proton of the H atom sticking out and is very vulnerable to be knocked off leaving the Cl atom a negatively charged ion (Cl-).

Halogen ions, like Cl-, bond with metal ions, like Na+, forming **salts**, like NaCl. The negative charge of the Cl- ions attracts the positive charge of the metal ions and the Cl- ion penetrates the opened part of the Na+ metal ion forming the table salt NaCl.

Teflon is a chain of carbon atoms covered by a coating of F atoms instead of H atoms as is the case in hydrocarbons. Things do not stick to Teflon because the Fatoms fit over the otherwise sharp unpaired electrons of the carbon chain like smooth snap buttons over barbs in a barbed wire. The smooth outward facing side of the F atom with its paired electrons so optimally and tightly fit that it leaves no empty space or no slightly sticking out electrons. This results in a very smooth and thus non sticky surface.

When these Teflon strands are woven into a sheet with micro pores, a textile called *Gore-Tex* is formed. The fabric is as smooth and impervious to water molecules like a mirror is to light. At the same time, its pores allow individual water vapour molecules in the air called humidity to pass through. It is impervious to water because it has a very smooth surface that does not tear and break up water droplets that are held together in a drop by the special bonds called hydrogen bonds. **Hydrogen bonds** are attractions between atoms due to the slightly positive charge of the protruding protons of the hydrogen of one molecule which are attracted to the slightly negative charge of the oxygen of another molecule.

Chlorine reacts with water to form *disinfectants* and *bleaches*. Chlorine reacts with water producing hydrochloric acid and oxygen which kill most bacteria. The oxygen reacts and destroys portions of molecules that absorb light of specific wavelengths causing colours to be bleached white.

Fluorine in water and toothpaste kill bacteria that produce cavity forming acids. All of the halogens and their compounds are very poisonous because of their activity. Because of their activity, they are not found free in nature, but are always combined in compounds. When freed by man's industries, they readily combine with O3 in the air destroying them. Ozone (O3) molecules are like protective umbrellas blocking harmful radiations from disturbing delicate forms such as life.

Ionic compounds and molecules

Atoms group together in 2 main types of stable formations; *molecules* and *ionic compounds*. Unpaired electrons of one atom bond with unpaired electrons of another atom forming a *molecule*. When the molecule has a charge due to a proton without a moderating electron, or an electron without a moderating proton, then the grouping of atoms is called an *ionic compound*.

When an electron is detached from the atom, the atom forms a positive ion of the element and exposes the positive charge of the lone proton left behind. When the free electron pairs with an unpaired electron of an atom it becomes mechanically attached like a snap mechanism. With the snout outwards it makes the atom it joined into an ion with a negative charge.

Ions of opposite charge attract and bond to each other like magnets using their EM fields. These bonds are called *ionic bonds* and cause compounds called ionic compounds. Ionic compounds like salts form brittle stable solids that when dissolved in water, break apart into charged ions called electrolytes that carry an electric current. They are necessary to sustain life. Ionic compounds are generally large and make up the hard face of our world in its rocks and salts.

Atoms can also bond mechanically when their shapes allow it. Atoms with unpaired electrons that extend out can be paired with unpaired extending electrons of other atoms. These bonds are called *covalent bonds* and atoms bound this way are called molecules. Molecular compounds are generally small and make up the soft face of our world in its rivers, air and life.

Acids, Bases, Salts and Soaps

Acids and bases are compounds that when dissolved in water break apart to form charged ions. Organic acids have a head made up of **COOH.** Inorganic acids are just a Halogen type atom with a bonding hydrogen like *HCl.* Acids easily lose a proton from the tightly held H atom that has its proton extending outward. The freed proton causes the remaining acid molecule with the lone electron to exhibit a charge and become an negative ion.

Organic acids are formed by oxidizing alcohol in a controlled way, just like alcohol is formed by oxidizing hydrocarbons in a controlled way. Acids are building blocks used in making the fats and proteins that make up animal life.

Bases are metallic compounds with an OH part that extends out sufficiently to be dislodged when dissolved in water. The OH is tightly bonded with the unpaired electron of the metal so that when it is lodged from the metal, it takes the unpaired electron with it leaving the metal with a lone proton. The positive ion of the base bonds with the negative ion of the acid to form an ionic bipolar compound called a salt. One end of the salt is acquired from the acid and the other end is acquired from the base.

The H+ from the acid and the OH- from the base form the neutral and stable molecule water (H2O).

When organic acids such as those derived from fats react with a metal base, *soaps* are produced. This can be likened to butter and cheese curdling out of the milk when an acid is added. Soap's functionality can be attributed to its shape. Like a salt, it is bipolar. One end acquired from the fat dissolves readily in fat, while the other end acquired from the metal of the base dissolves readily in water.

The strength of acids and bases are calibrated on a scale from 0pH to 14pH. Pure water has a pH of 7. When an acid is dissolved in water, the pH of the solution becomes less than 7. When a base is dissolved in water, the pH of the solution becomes greater than 7.

A strong solution of HCl has a pH of 0. The pH values of various solutions are listed below. Gastric acid (1), vinegar and lemon juice (2), orange juice (3), tomato juice (4), black coffee(5), urine (6), water (7), sea water (8), baking soda (9), milk of magnesia (10), ammonia (11), soapy water (12), bleach (13), a strong solution of NaOH (14).

Molecules of Minerals, Vegetables, and Animals

Plants based on *carbon* are soft, warm, and alive. Rocks based on *silicon* are hard, cold, and lifeless. Despite these clear differences, wood and rock have many similar properties and sometimes they are hard to visually differentiate them.

Animals live off plants and plants live off earth. Earth is ground up rock (sand) mixed with water and life forms most of which are too small to be seen. *Animals* and *plants* are based on carbon, oxygen and hydrogen atoms. *Rocks* are based on silicon, oxygen and metal atoms.

As we have seen, carbon and silicon atoms have a very similar shape accounting for the close relationship between minerals, vegetables, and animals.

Representing molecules

Molecules can be represented in various ways. Balls and sticks are traditionally used. Balls and sticks show the shape of molecules 3 dimentionally . The disadvantage of balls ans sticks is that they do not clearly show the bonds. As well the balls get in the way as you can not easily see what is behind the balls in larger crystals. Drawings of the molecular formulas show the bonds but fail to show the molecule`s shape. CHONX STIX shows both the shape and the bonds of molecules. They can be ordered by contacting the ather at andrewvecsey@hotmail.com . This booklet shows various molecules with the 3 ways of representing them.

Molecules in the air

Abundant Atmospheric Gases

Nitrogen, N_2 Oxygen, O_2

N2   O2

Greenhouse Gases

Water, H_2O Carbon Dioxide, CO_2 Methane, CH_4

H2O   CO2   CH4

Nitrous Oxide, N_2O Ozone, O_3

N2O   O3

Nitrogen, (N2) is a gas at room temperature and is colorless and odorless. Nitrogen is about 7th abundant element in our galaxy and the Solar System. It forms about 78% of Earth's atmosphere.

Oxygen (O2) is a highly reactive nonmetallic element and oxidizing agent that readily forms compounds (notably oxides) with most elements. By mass, oxygen is the 3rd most abundant element in the universe, after hydrogen and helium. At STP, 2 atoms of the element bind to form dioxygen with the formula O2. It is a diatomic gas that is colorless, odorless, and tasteless and forms 20% of air in the atmosphere. Many major classes of organic molecules in living organisms, such as proteins, nucleic acids, carbohydrates, and fats, contain oxygen, as do the major inorganic compounds that are constituents of animal shells, teeth, and bone. Most of the mass of living organisms is oxygen as it is a part of water, the major constituent of lifeforms. Water forms 60% of human body mass.

Ozone (O3) is a pale blue gas with a distinctively pungent smell. It is much less stable than O2 breaking down in the lower atmosphere to normal dioxygen. Ozone is formed from dioxygen by the action of ultraviolet light and also atmospheric electrical discharges.
Ozone's odor is sharp, reminiscent of chlorine, and detectable by many people at concentrations of as little as 10 ppb in air. Ozone is a far more powerful oxidant than dioxygen and has many Industrial and consumer applications related to oxidation. The ozone layer in the stratosphere prevents damaging ultraviolet light from reaching the Earth's surface.

Water (H2O) covers 71% of the Earth's surface. It is vital for all known forms of life. On Earth, 96.5% of the planet's water is found in seas and oceans, 1.7% in groundwater, 1.7% in glaciers and the ice caps of Antarctica and Greenland, a small fraction in other large water bodies, and 0.001% in the air as vapor, clouds (formed of solid and liquid water particles suspended in air), and precipitation. Only 2.5% of the Earth's water is fresh water, and 98.8% of that water is in ice and groundwater. Less than 0.3% of all freshwater is in rivers, lakes, and the atmosphere, and an even smaller amount of the Earth's freshwater (0.003%) is contained within biological bodies and manufactured products.

Carbon dioxide (CO2) is a gas at standard temperature and pressure and exists in Earth's atmosphere in this state, as a trace gas. As a solid, it is called dry ice. Plants, algae, and cyanobacteria use light energy to photosynthesize carbohydrate from CO2 and water, with O2 produced as a waste product. CO2 is produced during the processes of decay of organic materials and the fermentation of sugars in beer and winemaking. It is produced by combustion of wood, carbohydrates and major carbon- and hydrocarbon-rich fossil fuels such as coal, peat, petroleum and natural gas.

:C≡O:
112.8 pm

Carbon monoxide (CO) is a colorless, odorless, and tasteless gas that is slightly less dense than air. It is toxic to humans and animals when encountered in higher concentrations. CO is produced from the partial oxidation of carbon-containing compounds; it forms when there is not enough oxygen to produce CO_2, such as when operating a stove or an internal combustion engine in an enclosed space.

Like a man making love.

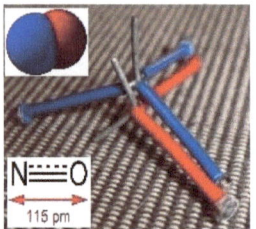

Nitric oxide (NO), or nitrogen oxide, also known as nitrogen monoxide, in mammals including humans is an important cellular signaling molecule involved in many physiological and pathological processes. Nitric oxide is rapidly oxidised in air to nitrogen dioxide.

Like 2 men making love.

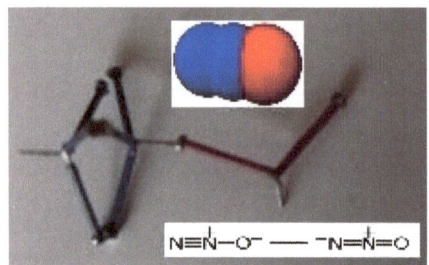

Nitrous oxide (N2O) is an anaesthetic and greenhouse gas used in surgery and dentistry for its effects to make you unconscious (anesthetic) and relieve pain (analgesic). It is known as laughing gas due to the euphoric effects of inhaling it.

Like trapeze artists.

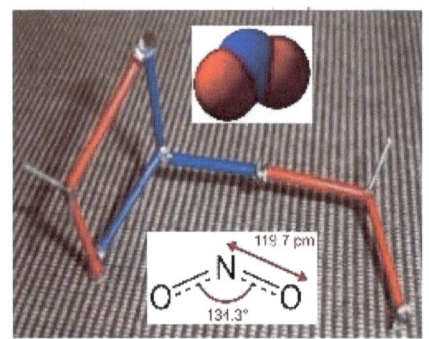

Nitrogen dioxide (NO2) is a brown toxic gas and a major air pollutant.

Like trapeze artists.

Molecules of fuel

Fuel molecules are chains of carbon containing H.

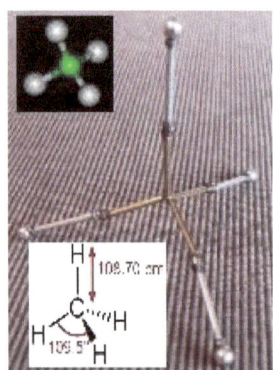

Methane (CH4) is the simplest hydrocarbon. It is a gas with a chemical formula of CH_4. It forms 80% of **natural earth gas**. Methane results from the decomposition of certain organic matters in the absence of oxygen.

Like a mother with 4 children hanging on.

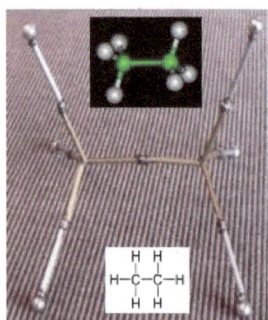

Ethane is a chemical compound with chemical formula C_2H_6, structural formula CH_3-CH_3.

Like 2 mothers with 3 children.

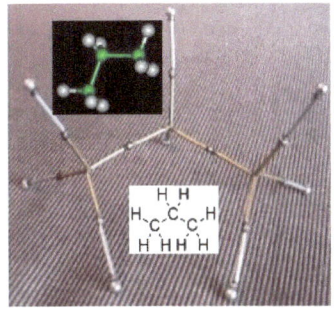

Propane (C3H3) is a 3-carbon alkane. It is sometimes derived from other petroleum products during oil or natural gas processing. When commonly sold as fuel, it is also known as liquified petroleum gas (LPG or LP gas) and is a mixture of propane with smaller amounts of propylene, butane and butylene, plus an ethyl mercaptan odorant to allow the normally odorless propane to be smelled. It is used as fuel in cooking on many barbecues and portable stoves and in motor vehicles. Propane powers some buses, forklifts, and taxies and is used for heat and cooking in recreational vehicles and campers. In many rural areas of the US, propane is also used in furnaces, water heaters, laundry driers, and other heat-producing appliances. Propane combustion is much cleaner than gasoline, though not as clean as natural earth gas.

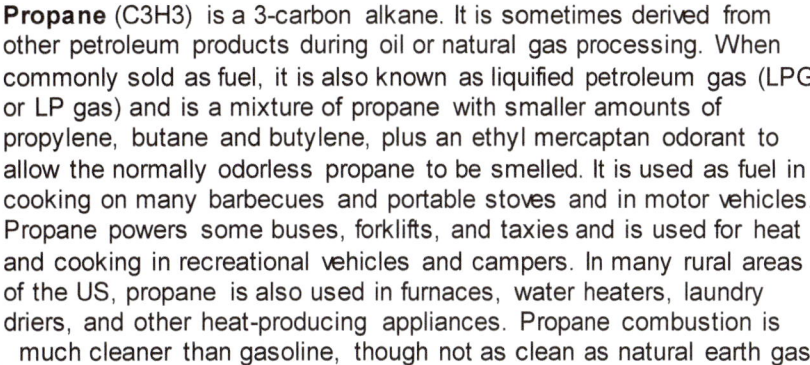

Butane is the unbranched alkane with four carbon atoms, $CH_3CH_2CH_2CH_3$. Butanes are highly flammable, colorless, odorless, easily liquefied gases. Butane gas is sold bottled as a fuel for cooking and camping. When blended with Propane and other hydrocarbons, it is referred to commercially as LPG.

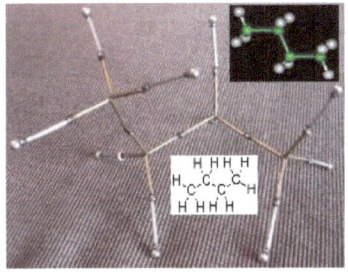

Pentane is one of three structural isomers with the molecular formula C_5H_{12}, the others being isopentane and neopentane. It is mainly a fuel and a solvent. The conformation (shape) of pentane is linear, similar to that of butane, but one carbon atom longer.

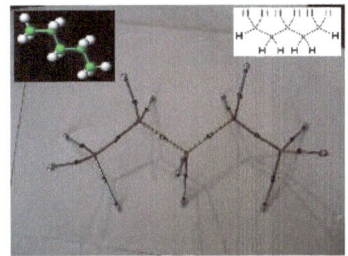

Hexane is an alkane hydrocarbon with the chemical formula $CH_3(CH_2)_4CH_3$. The "hex" prefix refers to its six carbons, while the "ane" ending indicates that its carbons are connected by single bonds. Hexane isomers are largely unreactive, and are frequently used as an inert solvent in organic reactions because they are very non-polar. They are also common constituents of gasoline and glues used for shoes, leather products and roofing. Additionally, it is used in solvents to extract oils for cooking and as a cleansing agent for shoe, furniture and textile manufacturing. Hexane is produced by the refining of crude oil.

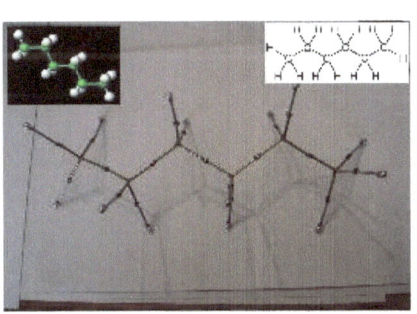

Heptane is the straight-chain alkane with the chemical formula $H_3C(CH_2)_5CH_3$ or C_7H_{16}.

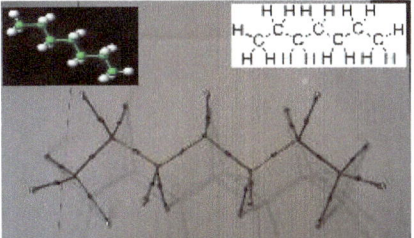

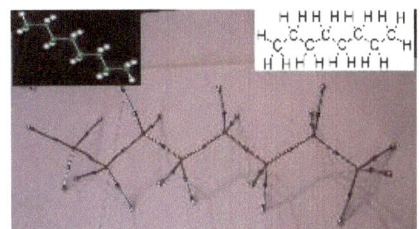

Octane (C8H18) is an alkane. It is an important constituent of petrol also called gasoline.

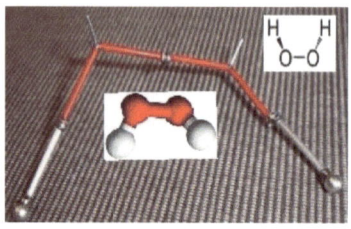

Hydrogen peroxide (H2O2) is a colorless liquid, slightly more viscous than water. It is a strong oxidizer and is used as a bleaching agent and disinfectant. Concentrated hydrogen peroxide, is used as a propellant for rockets. Organisms naturally produce trace quantities of hydrogen peroxide, most notably by a respiratory burst as part of the immune response. When it comes into contact with blood protein, it bubbles into water and oxygen gas.

Sugar molecules

Sugars are soluble carbohydrates- hydrocarbons with O. They are also called saccharides.

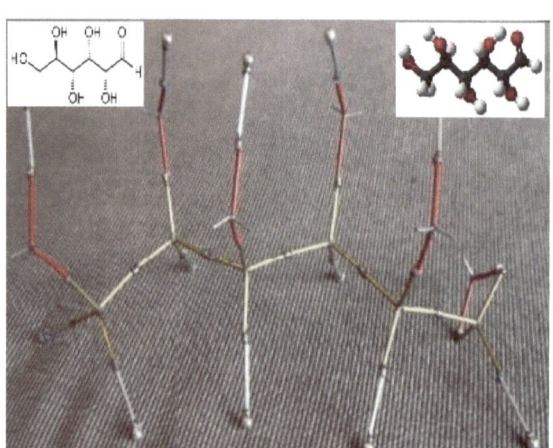

Glucose ($C_6H_{12}O_6$) also known as **blood sugar**, **dextrose** or **grape sugar** is a simple monosaccharide sugar. It is one of the most important carbohydrates and is used as a source of energy in animals and plants. Glucose is one of the main products of photosynthesis and starts respiration. It is absorbed directly into the bloodstream during digestion. **Starch**, **cellulose**, and **glycogen** are common glucose polymers (polysaccharides).

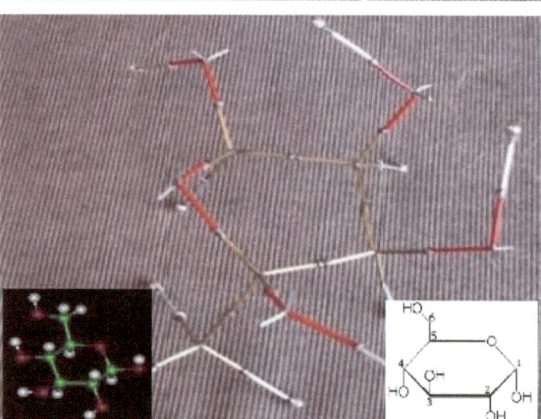

Glucose can exist in both a straight-chain and ring form.

Like a crab.

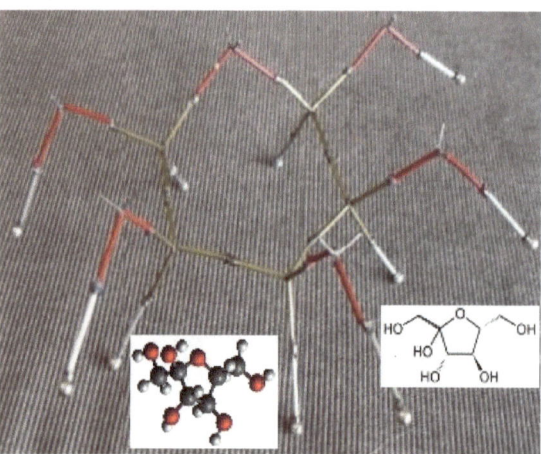

Fructose ($C_6H_{12}O_6$)or **fruit sugar**, is a simple ketonic monosaccharide found in many plants, where it is often bonded to glucose to form the disaccharide sucrose. Fructose, is the form of sugar found in fruit and honey. Fructose metabolise more slowly than cane sugar (sucrose) and is sweeter, so it has a smaller effect on blood-sugar levels.

Like a crab.

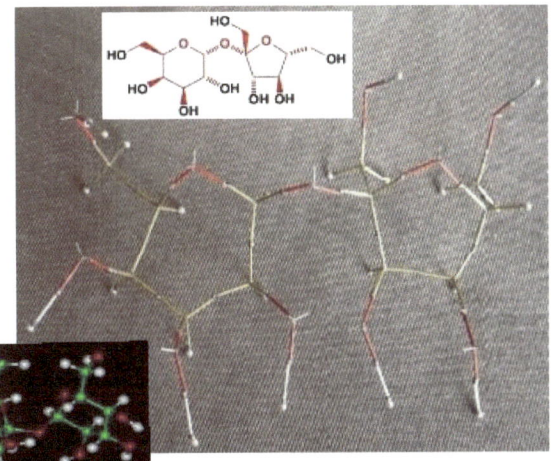

Sucrose also known as **table sugar** is a disaccharide (glucose + fructose) with the molecular formula $C_{12}H_{22}O_{11}$. It is best known for its role in human nutrition and is formed by plants but not by higher animals.

Like 2 crabs coupled together.

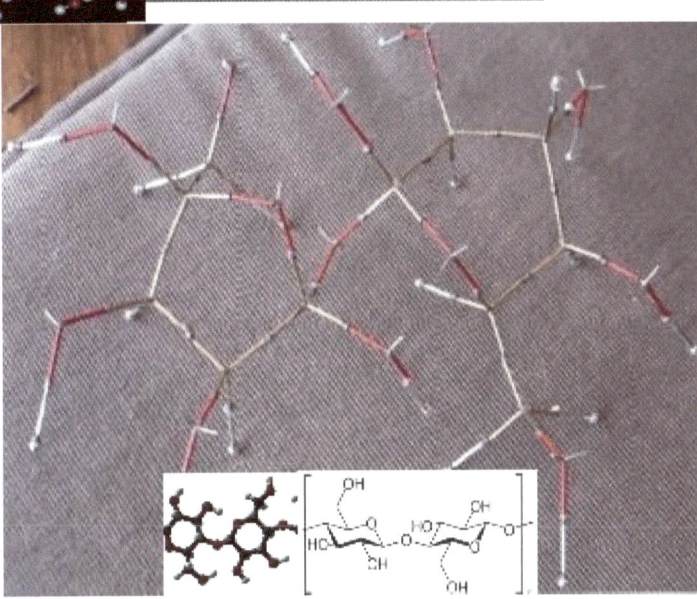

Cellulose (C6H10O5)n is a polysaccharide consisting of a linear chain of several hundred to many thousands of glucose units. Cellulose is an important structural component of the primary cell wall of green plants. Some species of bacteria secrete it to form biofilms. Cellulose is the most abundant organic polymer on Earth. The cellulose content of cotton fiber is 90%, that of wood is 40–50% and that of dried hemp is approximately 45%. Cellulose is mainly used to produce paperboard and paper. Smaller quantities are converted into a wide variety of derivative products such as cellophane and rayon. Conversion of cellulose from energy crops into biofuels such as cellulosic ethanol is under investigation as an alternative fuel source. Some animals, particularly ruminants and termites, can digest cellulose with the help of symbiotic micro-organisms that live in their guts.

Alcohols

An **alcohol** is any organic compound with the structure ROH where in which R is a satureated carbon chain.

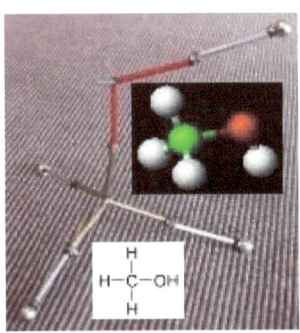

Methanol, also known as **wood alcohol** is a chemical compound with chemical formula CH_3OH. It is the simplest alcohol, and is a light, volatile, colourless, flammable, poisonous liquid with a distinctive odor that is somewhat milder and sweeter than ethanol (ethyl alcohol). Methanol ingested in large quantities is metabolized to formic acid which is poisonous to the central nervous system, and may cause blindness, coma, and death. At room temperature it is a polar liquid and is used as an antifreeze, solvent, and fuel.

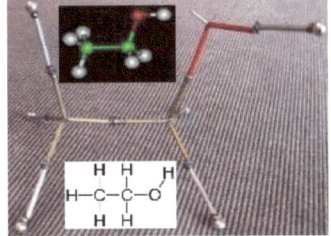

Ethanol also called **ethyl alcohol**, or **grain alcohol**, is a volatile, flammable, colorless liquid with the structural formula CH3CH2OH, often abbreviated as C2H5OH or C2H6O. Ethanol is the principal type of alcohol found in alcoholic beverages, produced by the fermentation of sugars by yeasts. It is a psychoactive drug and one of the oldest recreational drugs still used by humans. It is used in thermometers, as a solvent, as an antiseptic and as a fuel. When fully combusted its combustion products are only carbon dioxide and water. When fermented and oxidized, it forms vinegar, a diluted acetic acid.

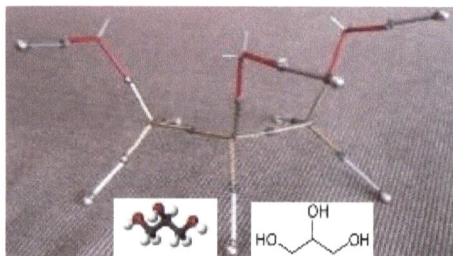

Glycerol, also called glycerin is a colorless, odorless, syrupy, sweet liquid, $C_3H_8O_3$, usually obtained by the saponification of natural fats and oils: used for sweetening and preserving food, in the manufacture of cosmetics, perfumes, inks, and certain glues and cements, as a solvent and automobile antifreeze, and in medicine in suppositories and skin emollients. The glycerol backbone is central to all fat molecules known as triglycerides.

Ethers

When the H at the end of an alcohol is replaced by a saturated carbon chain R`, **ethers** (ROR`) are produced.

Dimethyl ether (DME), (CH3OCH3), the simplest ether is a colourless gas that is a useful precursor to other organic compounds and an aerosol propellant. A potentially major use of DME is as substitute for propane used as fuel in household and industry. It is also a promising fuel in diesel engines, petrol engines and gas turbines.

Aldehydes

Aldehydes are dehydrogenated alcohols. They have the structure RCOH with the O forming a double bond with C and R is a saturated carbon chain. Many fragrances are aldehydes.

Formaldehyde (HCOH) is a gas at room temperature. It is colorless and has a characteristic pungent, irritating odor. Solutions of formaldehyde in water are used as disinfectants and for preservation of biological specimens. It is commonly used as nail varnish.

Ketones

When the H at the end of aldehydes is replaced by a saturated carbon chain R`, **ketones** (RCOR`) are produced.

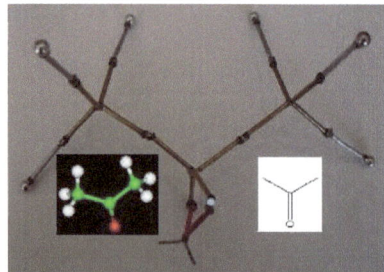

Acetone is the simplest representative of the ketones. Its chemical formula is $CH_3(CO)CH_3$. It is a colourless mobile flammable liquid with a pleasant, somewhat fruity odor. It is readily soluble in water, ethanol, ether, etc., and itself serves as an important solvent. The most familiar household use of acetone is as the active ingredient in nail-polish remover. Acetone is also used to make plastic, fibers, drugs, and other chemicals.

Acids, bases and salts

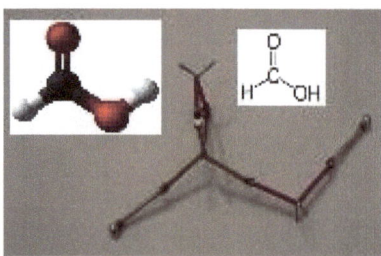

Formic acid (HCOOH) is the simplest carboxylic acid. It is an important intermediate in chemical synthesis and occurs naturally, most notably in ant venom. Formic acid has been shown to be an effective treatment against warts.

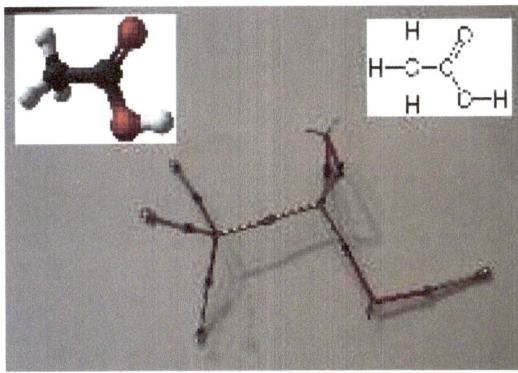

Acetic acid (CH₃COOH) is a colourless liquid. **Vinegar** is dilute acetic acid, often produced by fermentation and subsequent oxidation of ethanol. Acetic acid has a distinctive sour taste and pungent smell.

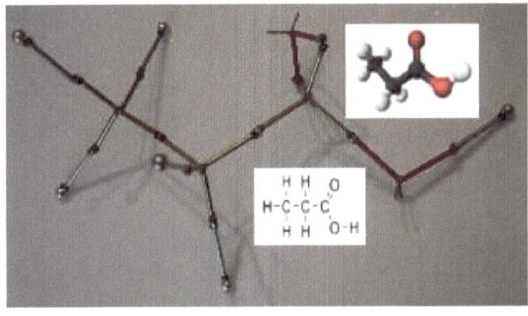

Propanoic acid (CH₃CH₂COOH) is a clear liquid with a pungent odor.

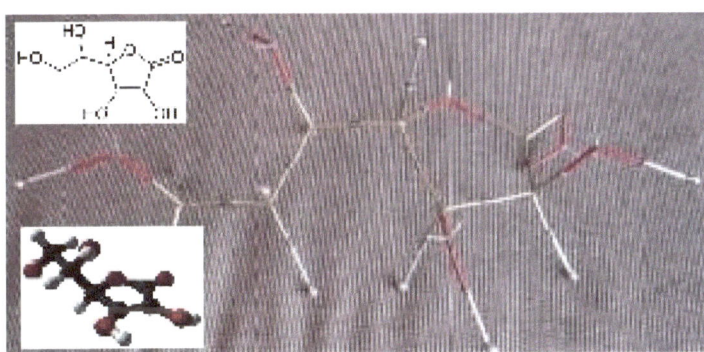

Vitamin C or **L**-ascorbic acid, or simply ascorbate, is an essential nutrient for humans and certain other animal species. Ascorbic acid is also widely used as a food additive, to prevent oxidation.

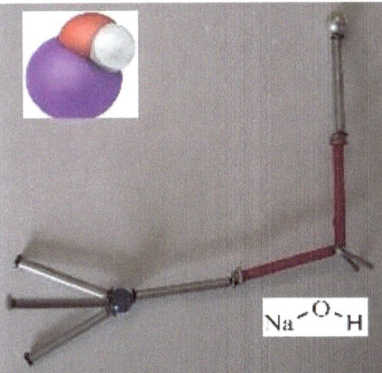

Sodium hydroxide, (NaOH) also known as **caustic soda**, or lye. It is a white solid and highly caustic metallic base and alkali salt which is available in pellets, flakes, granules, and as prepared solutions at a number of different concentrations.

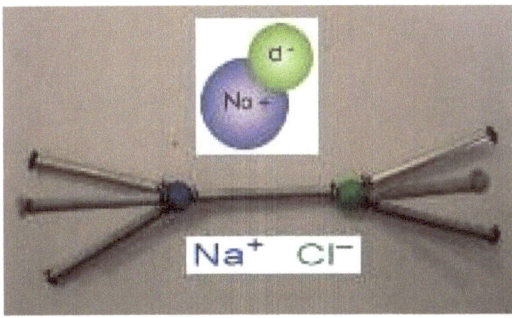

Hydrogen chloride (HCl) at room temperature, is a colorless gas, which forms white fumes of **hydrochloric acid** upon contact with atmospheric humidity.

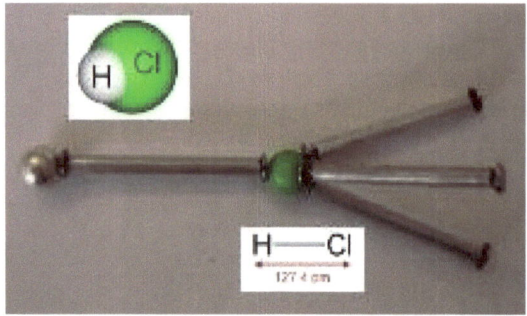

Sodium chloride, also known as **salt**, **common salt**, **table salt** or **halite**, is an ionic compound with the chemical formula NaCl, representing equal proportions of sodium and chlorine. Sodium chloride is the salt most responsible for the salinity of the ocean and of the extracellular fluid of many multicellular organisms. T has many uses from food preservative to deicing of roadways in sub-freezing weather.

Esters

Esters are generally derived from a carboxylic acid and an alcohol. Esters comprise most naturally occurring fats and oils, which are fatty acid esters of glycerol. Esters with low molecular weight are commonly used as fragrances and found in essential oils. Nitrate esters, such as nitroglycerin, are known for their explosive properties, while polyesters are important plastics.

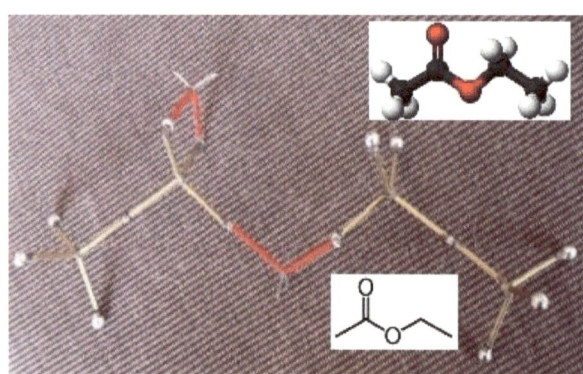

Ethyl acetate is a colorless liquid with a characteristic sweet smell (similar to pear drops) and is used in glues, nail polish removers, decaffeinating tea and coffee, and cigarettes.

Fat Molecule

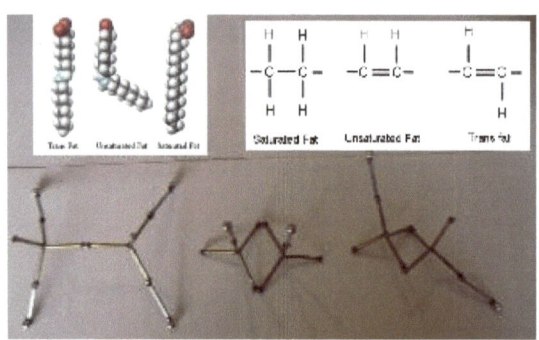

Fats are molecules that have an acid head with hydrocarbon chains. If the chain has no double bonds, the fats are called "saturated". With one or more double or triple bonds, they are called "unsaturated". Trans fats are unsaturated fats made in the laboratory that are twisted. When the fat molecule is liquid, it is called **oil**.

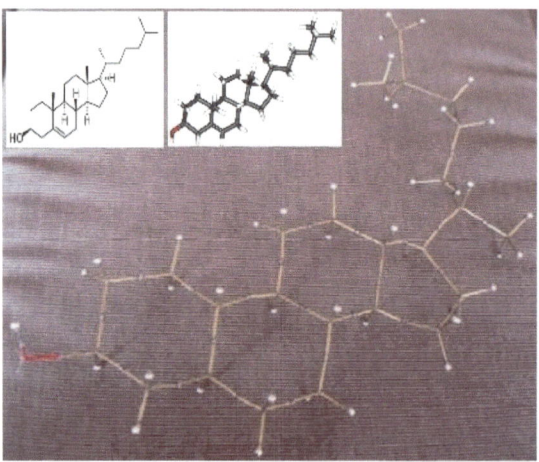

Cholesterol, is an essential structural component of animal cell membranes that is required to maintain both membrane structural integrity and fluidity. Cholesterol enables animal cells to (a) not need a cell wall (like plants & bacteria) to protect membrane integrity/cell-viability and thus be able to (b) change shape and (c) move about (unlike bacteria and plant cells which are restricted by their cell walls). In addition to its importance within cells, cholesterol also serves as a precursor for the biosynthesis of steroid hormones, bile acids, and vitamin D.

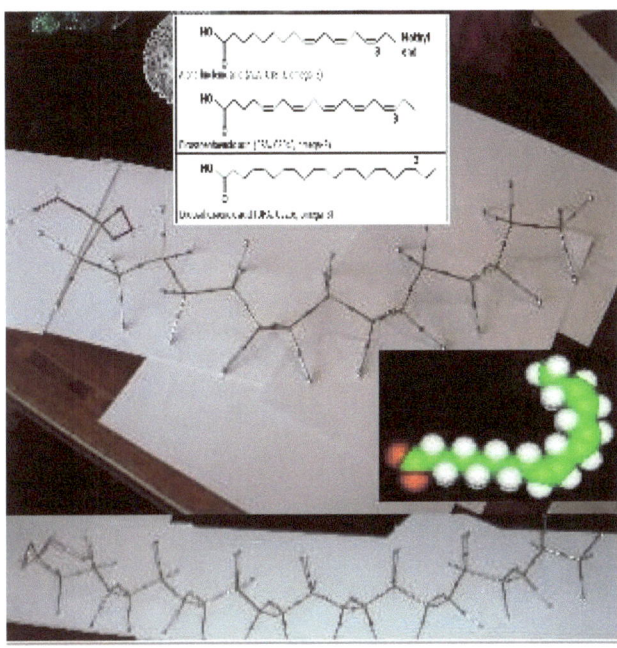

Omega-3 fatty acids are polyunsaturated fatty acids with a double bond (C=C) at the third carbon atom from the end of the carbon chain. The fatty acids have two ends, the carboxylic acid (-COOH) end, which is considered the beginning of the chain, thus "alpha", and the methyl (CH_3) end, which is considered the "tail" of the chain, thus "omega."

The 3 types of omega-3 fatty acids involved in human physiology are ALA (found in plant oils), EPA, and DHA (both commonly found in marine oils). In foods exposed to air, sunlight and high temperatures, unsaturated fatty acids are vulnerable to oxidation and rancidity. This makes them ideal components of paints.

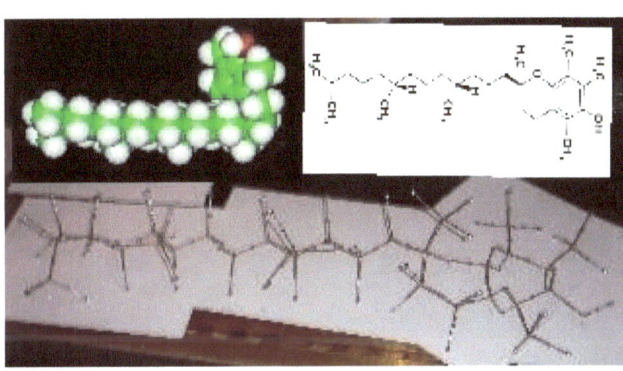

Tocopherol, or **Vitamin E**, is a fat-soluble vitamin in eight forms that is an important antioxidant.

Protein molecules

Proteins are molecules with an acid head and a carbon skeleton tail containing H and N atoms. Skin, muscle and bone are made from proteins.

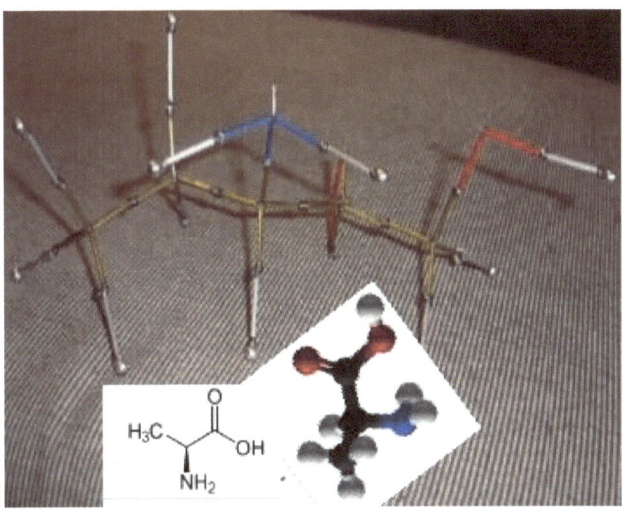

Alanine ($CH_3CH(NH_2)COOH$) is one of the 20 amino acids encoded by the genetic code.

Solvent molecules

Water is also a good solvent due to its polarity. The solvent properties of water are vital in biology, because many biochemical reactions take place only within aqueous solutions (e.g., reactions in the cytoplasm and blood). In addition, water is used to transport biological molecules.

The strong hydrogen bonds give water a high cohesiveness and, consequently, surface tension. This is evident when small quantities of water are put onto a nonsoluble surface and the water stays together as drops. This feature is important when water is carried through xylem up stems in plants; the strong intermolecular attractions hold the water column together, and prevent tension caused by transpiration pull. Other liquids with lower surface tension would have a higher tendency to "rip", forming vacuum or air pockets. Up to 10 meters of water can be sucked up by a vacuum pump.

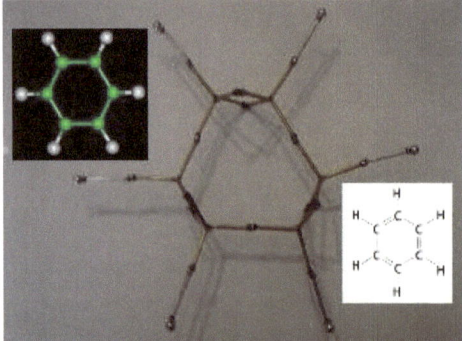

Benzene (C_6H_6) is a colorless, flammable, aromatic hydrocarbon, that is a known carcinogen. It is used in the creation of drugs, plastics, gasoline, synthetic rubber, napalm and dyes.

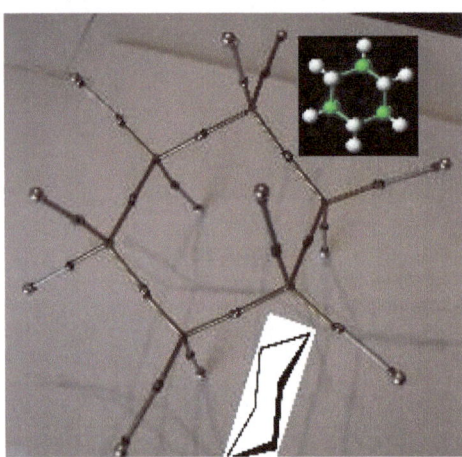

Cyclohexane is a cycloalkane with the molecular formula C_6H_{12}. Cyclohexane is used as a nonpolar solvent for the chemical industry, and also as a raw material for the industrial production of adipic acid and caprolactam, both of which are intermediates used in the production of nylon. On an industrial scale, cyclohexane is produced by reacting benzene with hydrogen.

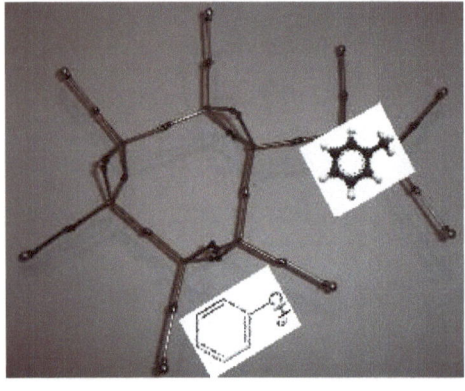

Toluene is a clear, water-insoluble liquid with the typical smell of paint thinners. It is widely used as a solvent. Like other solvents, toluene is sometimes also used as an inhalant drug for its intoxicating properties; however, inhaling toluene has potential to cause severe neurological harm.

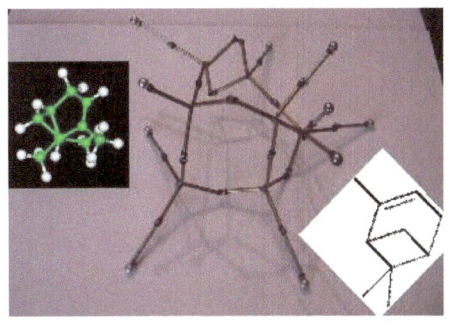

Turpentine is a fluid obtained by the distillation of resin obtained from live trees, mainly pines. It is mainly used as a solvent and as a source of materials for organic synthesis. It contributes to the smell of the forest and also to the smell of some fruits.

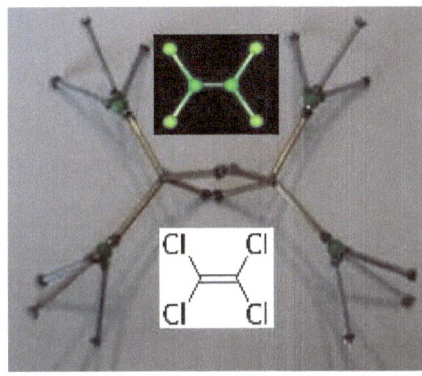

Tetrachloroethylene ($Cl_2C=CCl_2$) dissolves most organic materials. It is the most widely used solvent in dry cleaning. It is also used to degrease metal parts in the automotive and other metalworking industries. It appears in a few consumer products including paint strippers and spot removers.

Molecules of color

Colour is produced by large molecules forming a ring with flat shapes resembling drum heads of **percussion instruments**. For these molecules to resonate, they have to be lit or hit by frequencies close to their resonant frequency. Their resonant frequency is the colour that we see. Chemicals called bleaches are known to whiten colours. They are corrosive and disturb the bonds that cause the colours by breaking them. *Like an iron flattening out creases in cloth.*

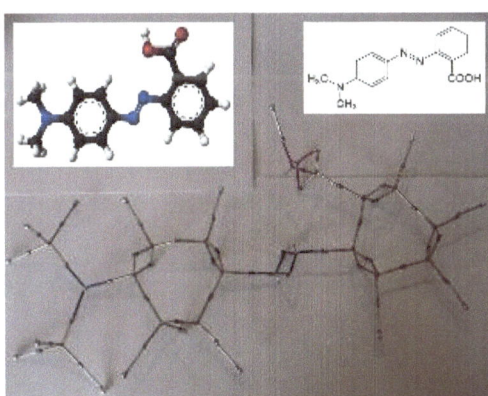

Methyl red is an indicator dye that turns red in acidic solutions.

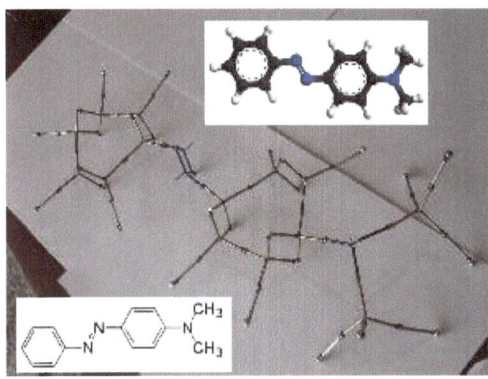

Methyl yellow, **Methyl yellow** is used as a pH indicator. In aqueous solution at low pH, methyl yellow appears red. Between pH 2.9 and 4.0, methyl yellow undergoes a transition, to become yellow above pH 4.0.

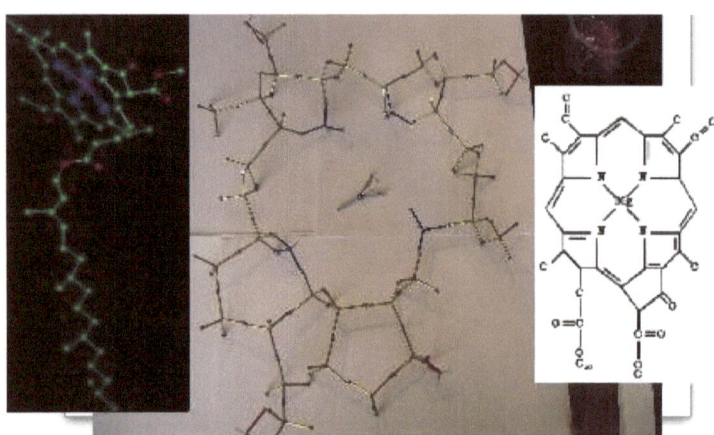

Chlorophyll is a green pigment found in most plants, algae, and cyanobacteria. Chlorophyll is vital for photosynthesis, which allows plants to obtain energy from light. Chlorophyll absorbs light and transfer that light energy by resonance energy transfer.

Like a kite attrackting lightning.

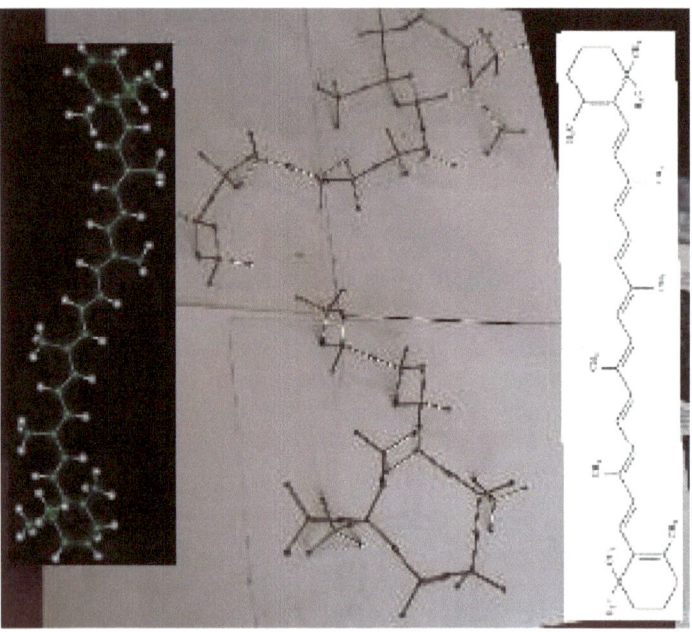

Carotene plays a crucial role as a photosynthetic pigment, important for photosynthesis. . It does not actively contribute in photosynthesis, but instead it transmits the energy it absorbs to chlorophyll and also plays a protective role for chlorophyll being a powerful antioxidant that protects organic molecules from being destroyed by oxidation.

Molecules of smell and taste

Smell and taste are produced by molecules forming chains resembling strings of **string instruments**. For these molecules to resonate, they have to be heated by frequencies close to their resonant frequency. Their resonant frequency is the smell and taste we perceive. We smell the volatile molecules and we taste the non volatile ones. Carbohydrates are *sweet*, acids are *sour*, salts are *salty* and bases are *bitter*.

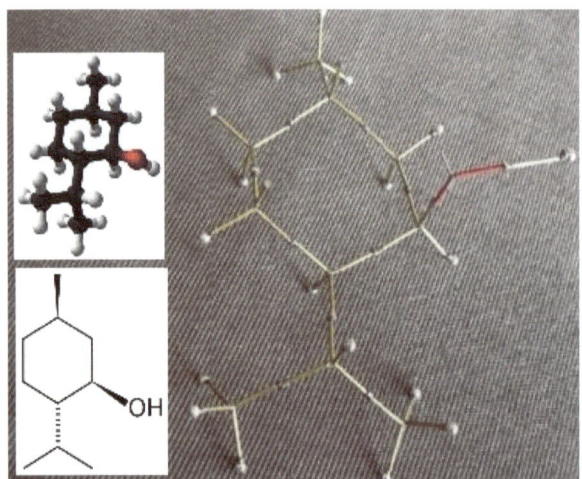

Menthol is a waxy, crystalline substance, clear or white in color, which is solid at room temperature and melts slightly above. Menthol has local anesthetic and counterirritant qualities, and it is widely used to relieve minor throat irritation.

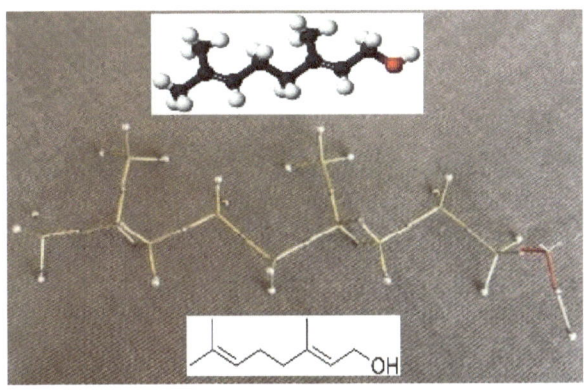

Geraniol has a rose-like scent and is commonly used in perfumes. It is used in flavors such as peach, raspberry, grapefruit, red apple, plum, lime, orange, lemon, watermelon, pineapple, and blueberry.

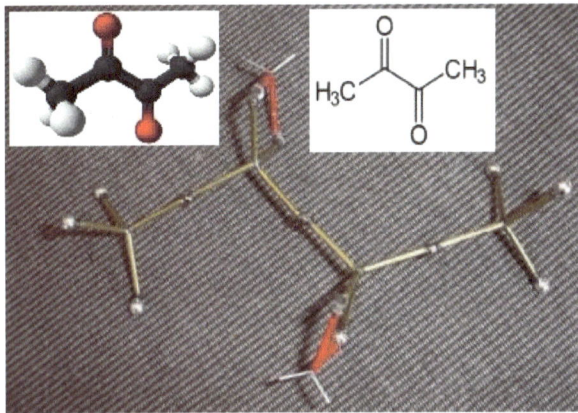

Linalool is a naturally occurring terpene alcohol found in many flowers and spice plants with many commercial applications, the majority of which based on its pleasant scent (floral, with a touch of spiciness).

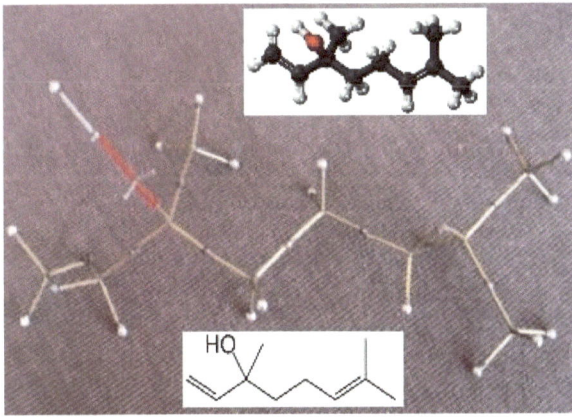

Diacetyl is a volatile, yellow/green liquid with an intensely buttery flavor. It occurs naturally in alcoholic beverages and is added to some foods to impart its buttery flavor.

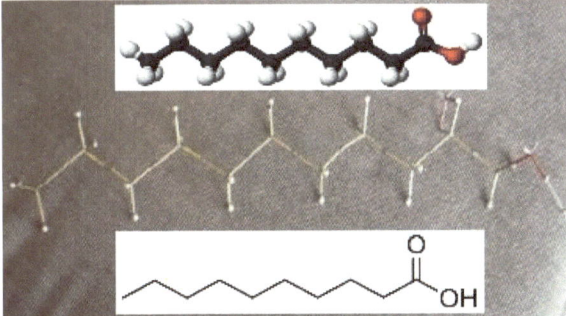

Decanoic acid, or **capric acid**, is a saturated fatty acid . It is found in the milk of various mammals and to a lesser extent in other animal fats. It smells like goats milk.

Butyric acid is found in milk, especially goat, sheep and buffalo milk, butter, Parmesan cheese, and as a product of anaerobic fermentation (including in the colon and as body odor). Butyric acid is present in, and is the main distinctive smell of, human vomit.

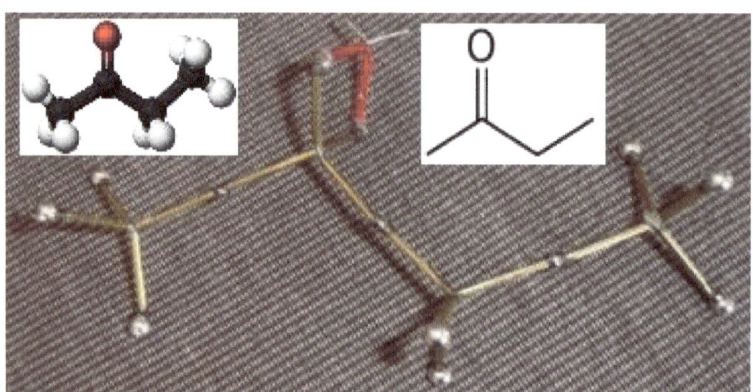

Butanone, is a colorless liquid ketone with a sharp, sweet odor reminiscent of butterscotch and acetone

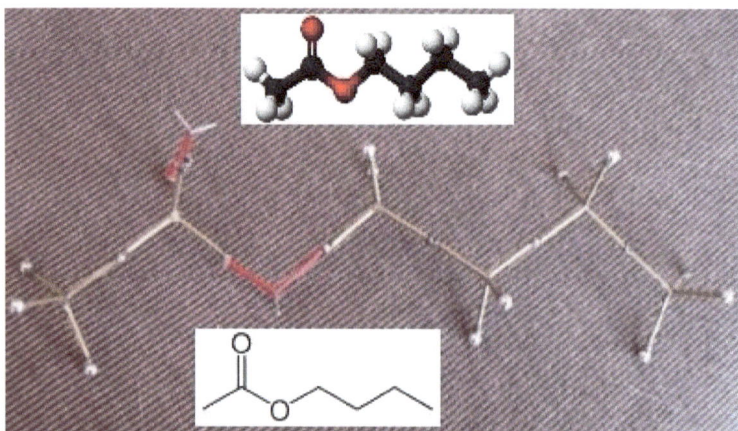

n-**Butyl acetate**, is found in many types of fruit, where along with other chemicals it imparts characteristic flavors and has a sweet smell of banana or apple. It is used as a synthetic fruit flavoring in foods such as candy, ice cream, cheeses, and baked goods.

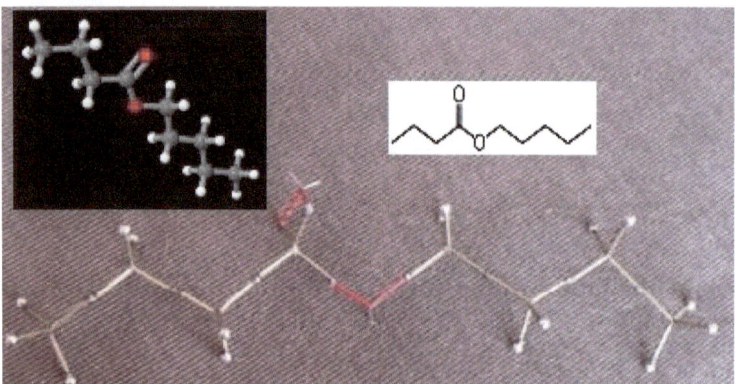

Amyl butyrate has a smell reminiscent of pear or apricot. This chemical is used as an additive in cigarettes.

Diallyl disulfide is derived from garlic and has a strong garlic odor. Highly diluted, it is used as a flavoring in food.

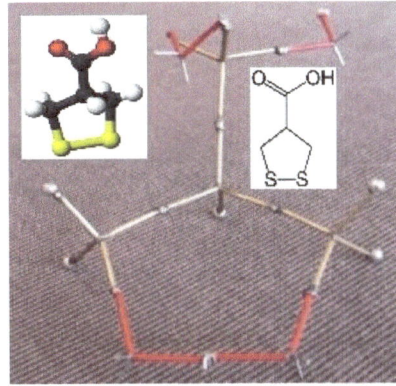

Asparagusic acid is present in the vegetable asparagus and it is the cause of its distinctive odor.

Molecules for emotions

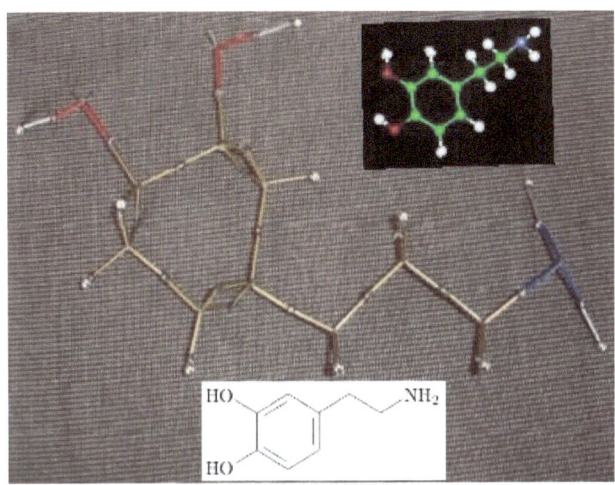

Dopamine ($C_6H_3(OH)_2$-CH_2-CH_2-NH_2) is a neurotransmitter in the brain. It is a hormone that is involved in the chemistry of pleasure. Although it functions normally to reward vital activities such as eating and sex, this same mechanism is also responsible for the craving connected with addiction to drugs, cocaine for example.

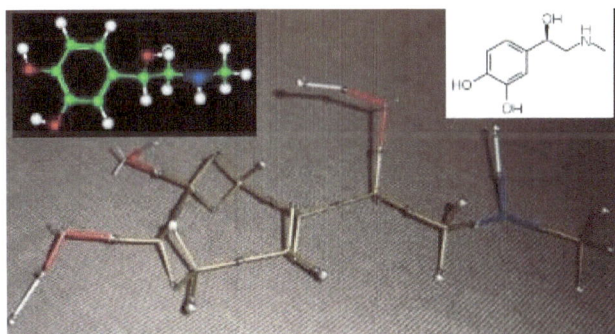

Epinephrine or adrenaline is a hormone when carried in the blood and a neurotransmitter when it is released across a neuronal synapse.

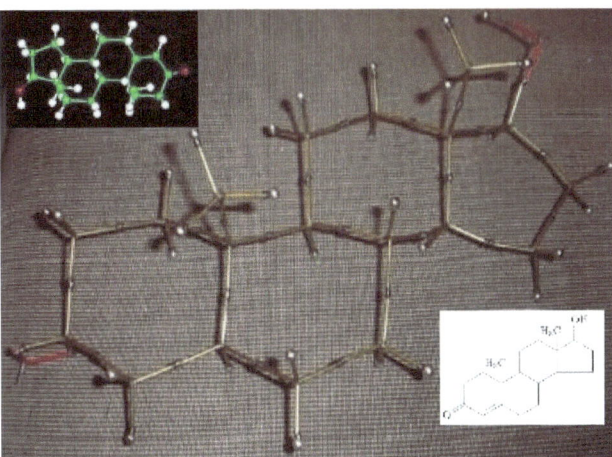

Testosterone, like other steroid hormones is derived from cholesterol. The largest amounts of testosterone are produced by testes

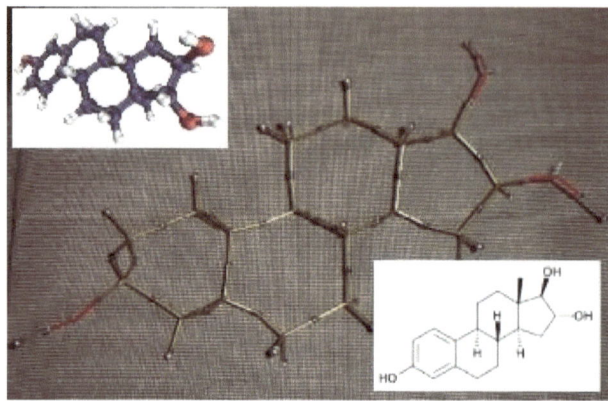

Estrogens are a group of steroid compounds, named for their importance in the oestrus cycle, functioning as the primary female sex hormone.

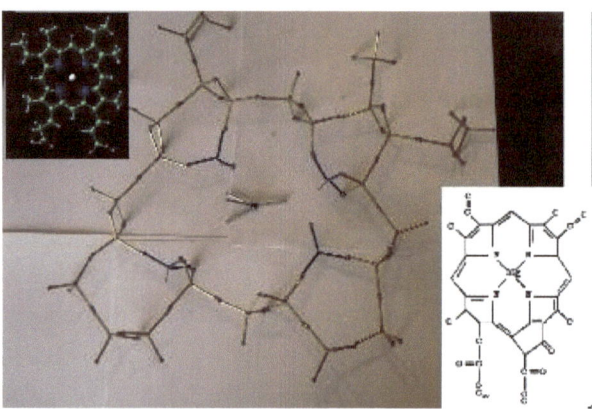

Hemoglobin, is the iron-containing oxygen-transport molecule that transports oxygen from the lungs or gills to the rest of the body, such as to the muscles, where it releases the oxygen.

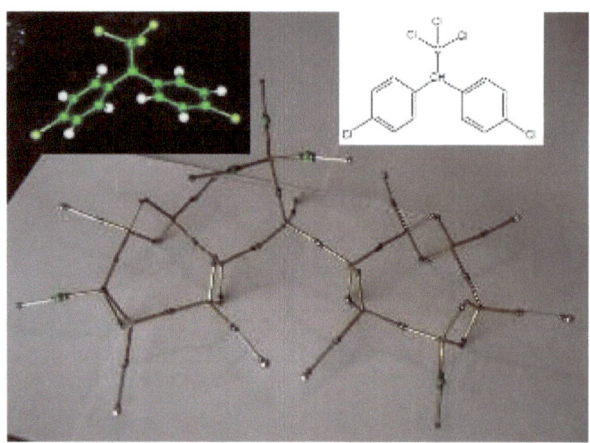

DDT $(ClC_6H_4)_2CH(CCl_3)$ is a colourless crystalline organochloride insecticide. It is very soluble in fats and most organic solvents and practically insoluble in water.

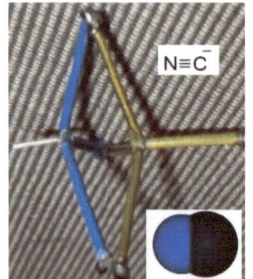

Cyanides

A **cyanide** is any chemical compound that contains monovalent combining group CN. This group, known as the **cyano group**, consists of a carbon atom triple-bonded to a nitrogen atom.

In inorganic cyanides, such as sodium cyanide, NaCN, this group is present as the negatively-charged polyatomic **cyanide ion** (CN^-); these compounds, which are regarded as salts of hydrocyanic acid, are highly toxic. Most cyanides are highly toxic.

Organic cyanides are usually called nitriles; in these, the CN group is linked to a carbon-containing group, such as methyl (CH_3) in methyl cyanide (acetonitrile).

Cyanides are employed in a number of chemical processes, including fumigation, case hardening of iron and steel, electroplating, and the concentration of ores. In nature, substances yielding cyanide are present in certain seeds, such as the pit of the wild cherry.

Cyanides are produced by certain bacteria, fungi, and algae and are found in a number of plants. Cyanides are found in substantial amounts in certain seeds and fruit stones, e.g., those of apricots, apples, and peaches. In plants, cyanides are usually bound to sugar molecules and defend the plant against herbivores.

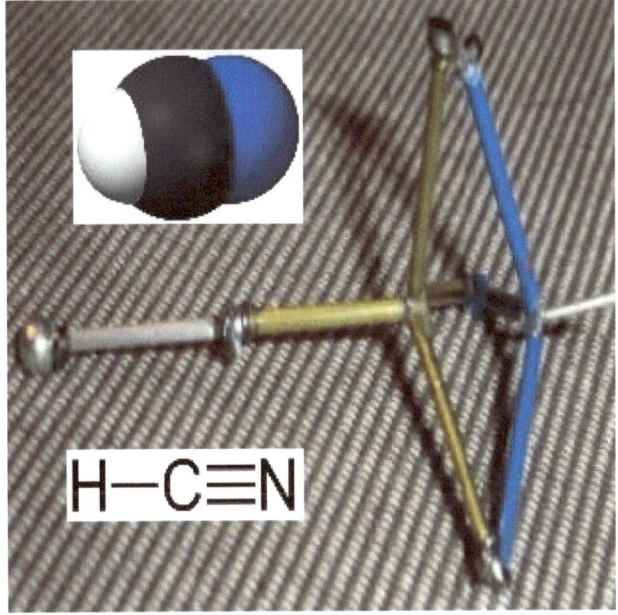

Hydrogen cyanide (HCN), sometimes called **prussic acid**, is a colorless, extremely poisonous liquid. It is a highly valuable precursor to many chemical compounds ranging from polymers to pharmaceuticals. It isused in the production of acrylic fibers, synthetic rubber, and plastics. Hydrogen cyanide was first isolated from a blue pigment (Prussian blue) which had been known from 1704 but whose structure was unknown.

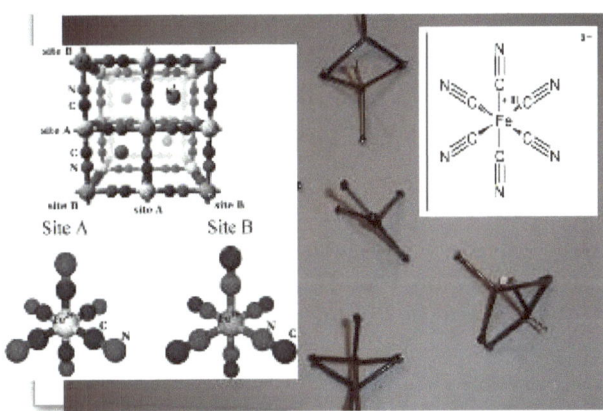

Prussian blue (Fe$_7$(CN) is a dark blue pigment. It was one of the first synthetic pigments. It is employed as a very fine colloidal dispersion, as the compound itself is not soluble in water. The pigment is used in paints, and it is the traditional "blue" in blueprints. In medicine, Prussian blue is used as an antidote for certain kinds of heavy metal poisoning.

Sodium cyanide (NaCN) is a white, water-soluble solid. When it is treated with acid, it forms the toxic gas hydrogen cyanide. Sodium cyanide is produced by treating hydrogen cyanide with sodium hydroxide: HCN + NaOH → NaCN + H$_2$O

Explosive molecules

Explosives are substances that contain a large amount of energy stored in chemical bonds. The energetic stability of the gaseous products and hence their generation comes from the formation of strongly bonded species like carbon monoxide, carbon dioxide, and (di)nitrogen, which contain strong double and triple bonds. Consequently, most commercial explosives are organic compounds containing -NO$_2$, -ONO$_2$ groups that, when detonated, release gases.

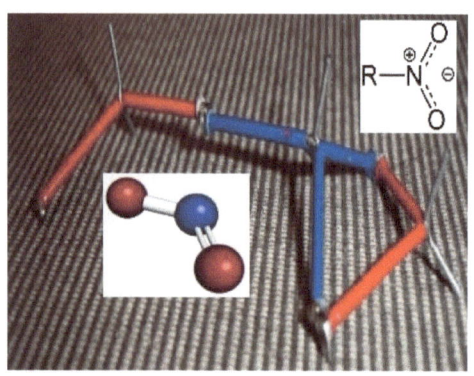

Nitro compounds are organic compounds that contain one or more **nitro** functional groups (–NO₂). They are often highly explosive, especially when the compound contains more than one nitro group and is impure.

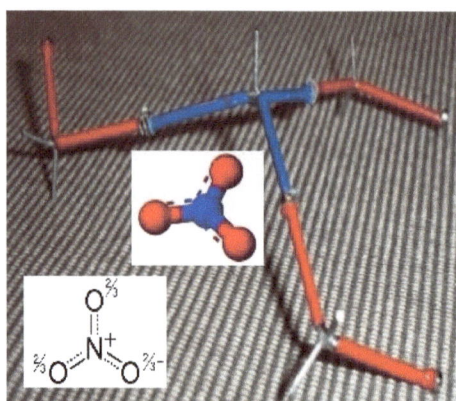

Nitrate is a polyatomic ion with the molecular formula NO_3^-. Nitrates also describe the organic functional group $RONO_2$. These nitrate esters are a specialized class of explosives.

Like a saw blade.

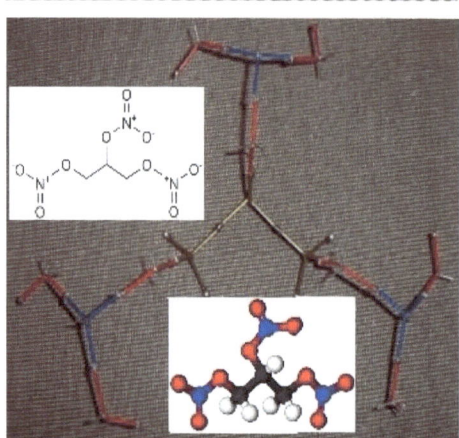

Nitroglycerin (NG), is a heavy, colorless, oily, explosive liquid most commonly produced by treating **glycerol** with **nitric acid**. It is an active ingredient in the manufacture of explosives, mostly dynamite, Detonation of nitroglycerin generates gases that would occupy more than 1,200 times the original volume at ordinary room temperature and pressure. Moreover, the heat liberated raises the temperature to about 5,000 °C.

Like a circular saw.

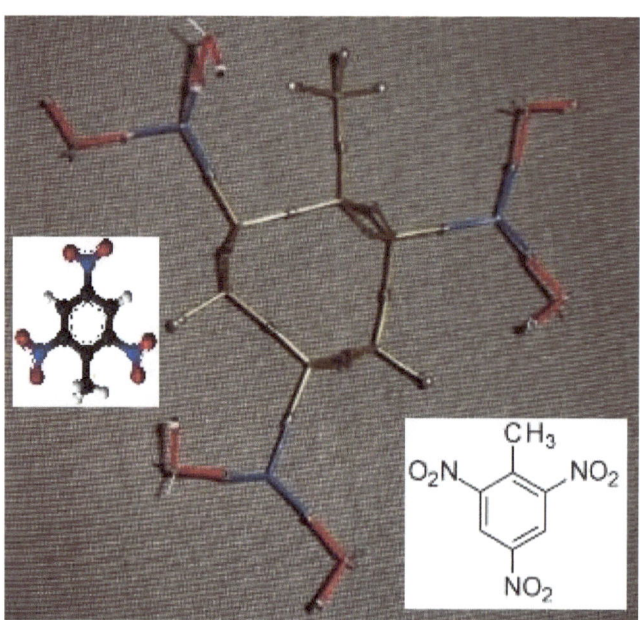

Trinitrotoluene (TNT), ($C_6H_2(NO_2)_3CH_3$) is a yellow-colored solid best known as a useful explosive material. TNT is one of the most commonly used explosives for military and industrial applications. It is valued partly because of its insensitivity to shock and friction, which reduces the risk of accidental detonation, compared to other more sensitive high explosives such as nitroglycerin.

Like a fighting crab.

Drug molecules

Stimulants

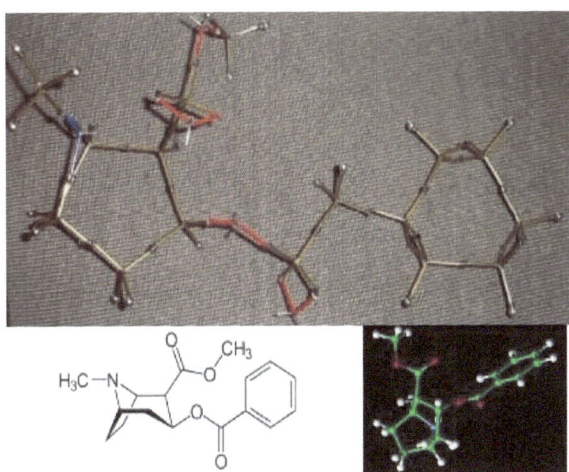

Cocain is a crystalline alkaloid that is obtained from the leaves of the coca plant. It is both a stimulant of the central nervous system and an appetite suppressant. Specifically, it is a dopamine reuptake inhibitor. It gives a feeling to what has been described as a euphoric sense of happiness and increased energy. It is most often used recreationally for this effect. For a thousand years, South American indigenous peoples have chewed the coca leaf that contains vital nutrients as well as numerous alkaloids, including cocaine. Cocaine in its purest form is a white, pearly product. Cocaine appearing in powder form is a salt. Its possession, cultivation, and distribution are illegal for non-medicinal and non-government sanctioned purposes in virtually all parts of the world.

Like a lizard.

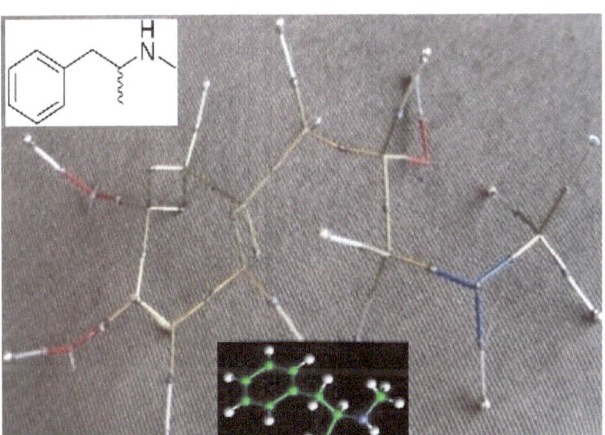

Methamphetamine has the chemical formula $C_{10}H_{15}N$. Since it causes euphoria and excitement, it is prone to abuse and addiction. Users may become obsessed or perform repetitive tasks such as cleaning, hand-washing, or assembling and disassembling objects. Withdrawal is characterized by excessive sleeping, eating and depression-like symptoms, often accompanied by anxiety and drug-craving.

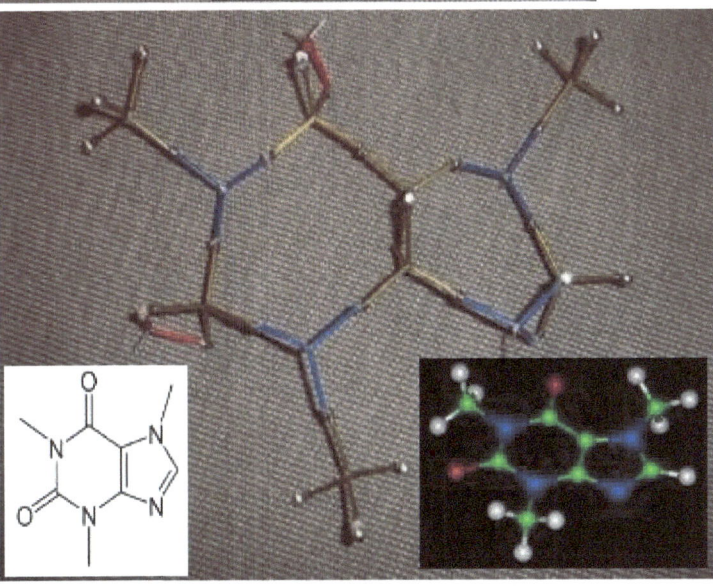

Caffeine is a stimulant drug. Caffeine is found in varying quantities in the beans, leaves, and fruit of over 60 plants, where it acts as a natural pesticide that paralyzes and kills certain insects feeding on the plants. It is most commonly consumed by humans in infusions extracted from the *beans* of the *coffee plant* and the *leaves* of the *tea bush*. Caffeine is a central nervous system stimulant, having the effect of temporarily warding off drowsiness and restoring alertness. Beverages containing caffeine, such as coffee, tea, soft drinks and energy drinks enjoy great popularity; caffeine is the world's most widely consumed psychoactive substance, but unlike most other psychoactive substances, it is legal and unregulated in nearly all jurisdictions. In North America, 90% of adults consume caffeine daily.

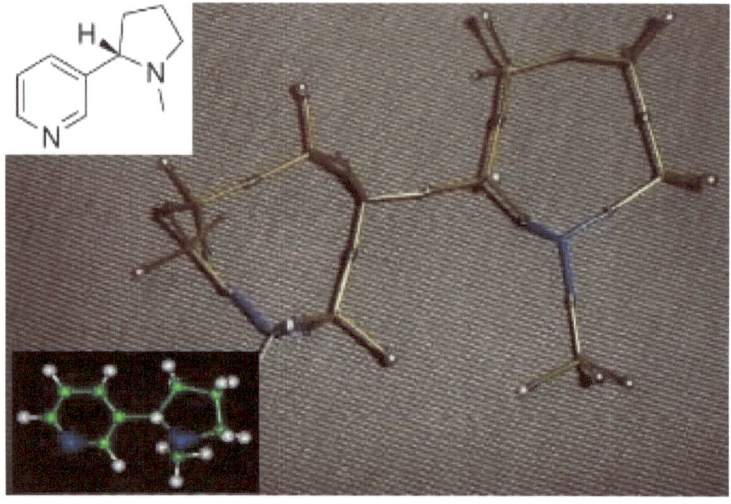

Nicotine is an alkaloid found in the nightshade family of plants, predominantly in tobacco and coca, and in lower quantities in tomato, potato, eggplant (aubergine), and green pepper. It functions as an antiherbivore chemical, being a potent neurotoxin with particular specificity to insects; therefore nicotine was widely used as an insecticide in the past, and currently nicotine derivatives such as imidacloprid continue to be widely used. In low concentrations, it acts as a stimulant in mammals and is one of the main factors responsible for the dependence-forming properties of tobacco smoking. According to the American Heart Association, "Nicotine addiction has historically been one of the hardest addictions to break."

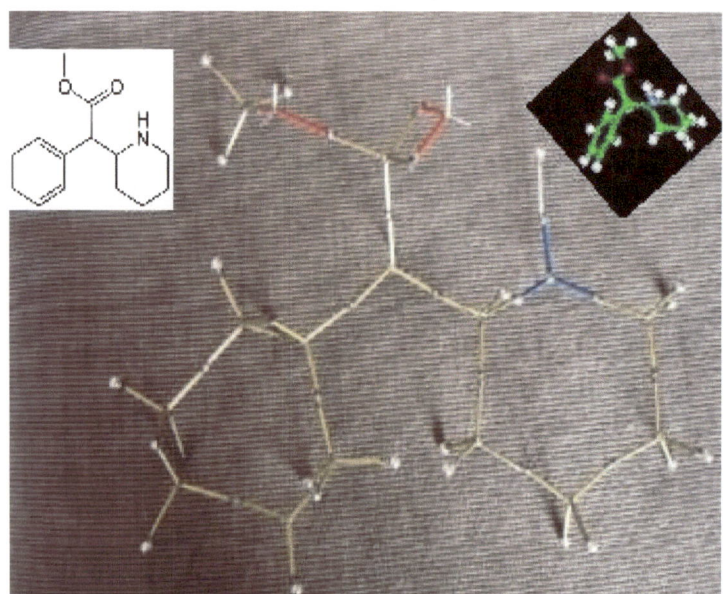

Methylphenidate (MPH) is most commonly known by the trademark name **Ritalin**. It is a prescription stimulant commonly used to treat Attention-deficit hyperactivity disorder, or ADHD. It is also one of the primary drugs used to treat the daytime drowsiness symptoms of narcolepsy and chronic fatigue syndrome. The drug is seeing early use to treat cancer-related fatigue.

Like Nicotine with a tail.

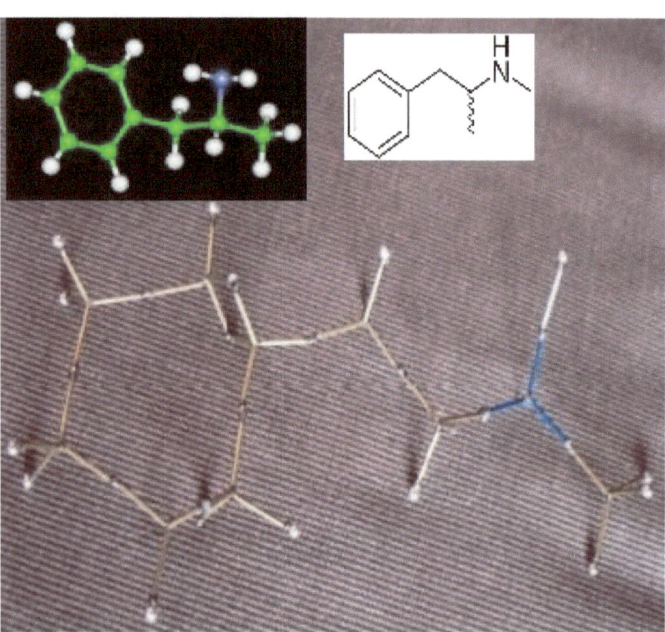

Amphetamine is a prescription stimulant commonly used to treat attention-deficit hyperactivity disorder (ADHD) in adults and children. It is also used to treat chronic fatigue syndrome. Initially it was more popularly used to diminish the appetite and to control weight. Brand names of the drugs that contain amphetamine include Adderall and Dexedrine. The drug is also used illegally as a recreational club drug and as a performance enhancer.

Like an ant.

Analgesics

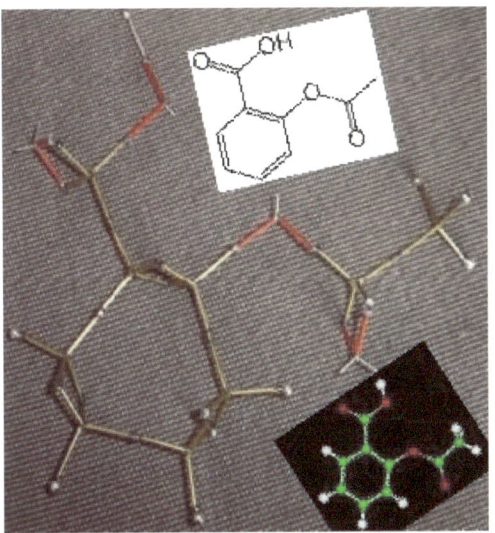

Aspirin is a brand name coined by the Bayer company of Germany for acetylsalicylic acid, part of the family of salicylates, often used as an analgesic, antipyretic, and anti-inflammatory.

Hippocrates, a Greek for whom the Hippocratic Oath is named, wrote in the 5th century BC about a bitter powder extracted from willow bark that could ease aches and pains and reduce fevers. This remedy is also mentioned in texts from ancient Sumeria, Egypt and Assyria. Native American Indians used it for headaches, fever, sore muscles, rheumatism, and chills.

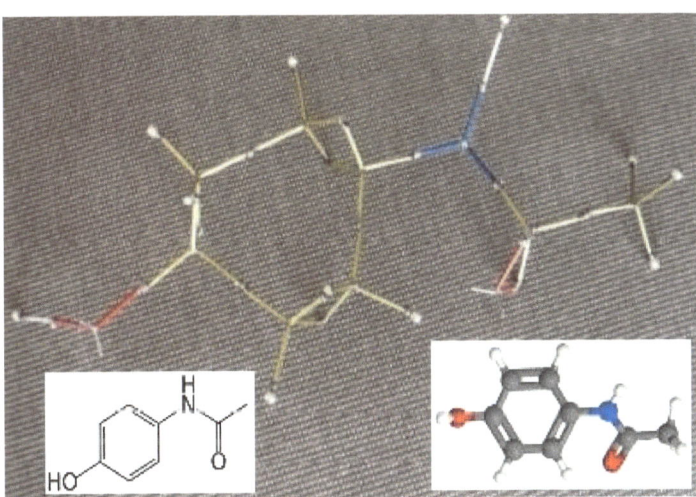

Paracetamol or **acetaminophen** (USAN), is the active metabolite of phenacetin, a so-called coal tar analgesic. It has analgesic and antipyretic properties.

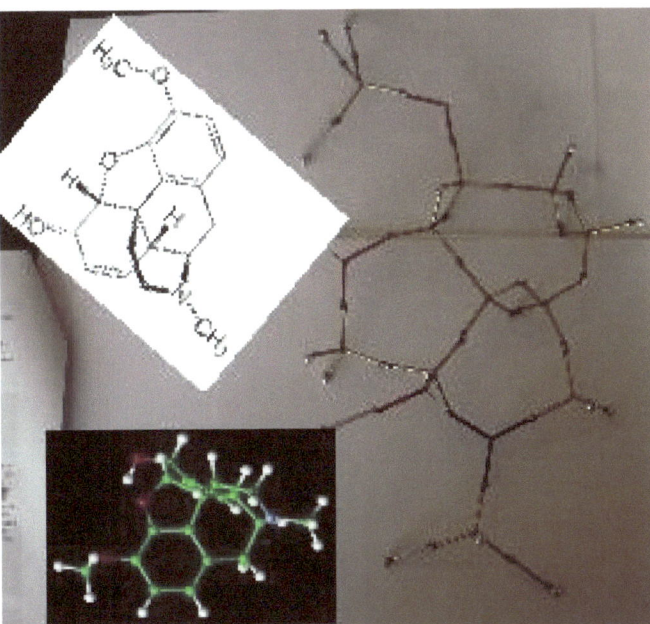

Codeine ($C_{18}H_{21}NO_3$ or **methylmorphine** is an alkaloid found in opium. It is an opiate used for its analgesic, antitussive and antidiarrheal properties. It is by far the most widely used opiate in the world and very likely most commonly used drug overall. It is about 10% of the strength of morphine. While codeine can be extracted from opium, most codeine is synthesized from morphine.

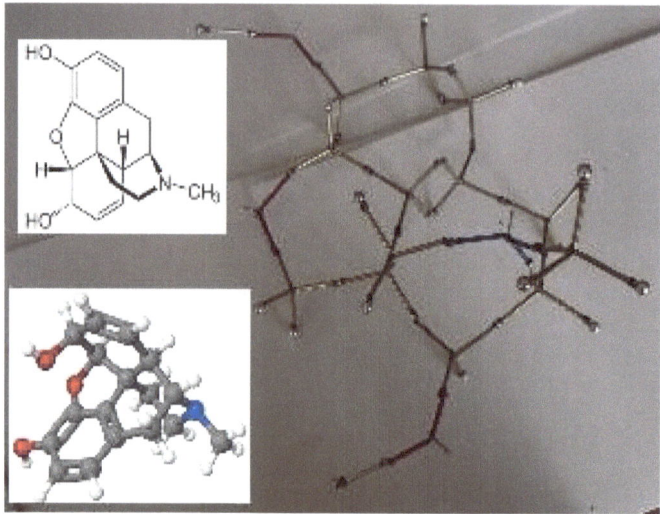

Morphine (C17H19O3) is a highly-potent opiate analgesic drug and is the principal active agent in opium. Like other opiates, e.g., diacetylmorphine (heroin), morphine acts directly on the central nervous system (CNS) to relieve pain. Studies done on the efficacy of various opioids have indicated that, in the management of severe pain, no other narcotic analgesic is more effective or superior to morphine. Morphine is highly addictive.

Like Codein with a dagger.

Amtibiotics

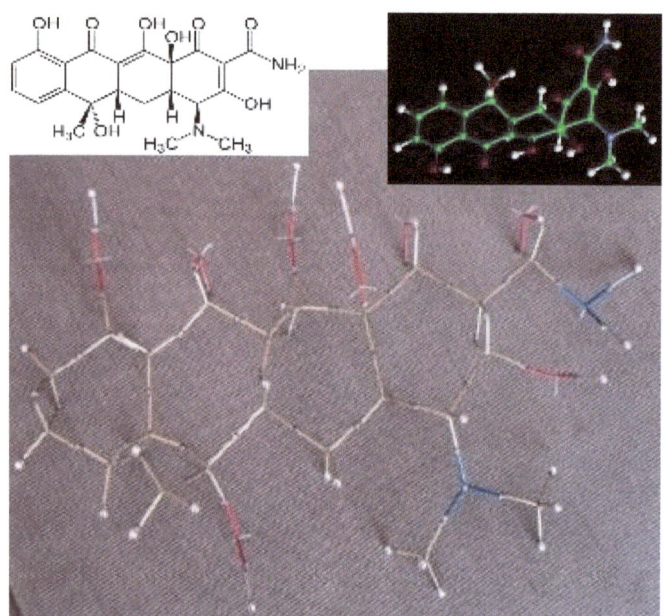

Tetracycline has a chemical formula of $C_{22}H_{24}N_2O_8$, and is used against many bacterial infections. It is commonly used to treat acne.

Sedatives

Barbiturates are drugs that act as central nervous system depressants, and by virtue of this they produce a wide spectrum of effects, from mild sedation to anesthesia. Some are also used as anticonvulsants. Barbiturates are derivatives of barbituric acid.

Like a bug.

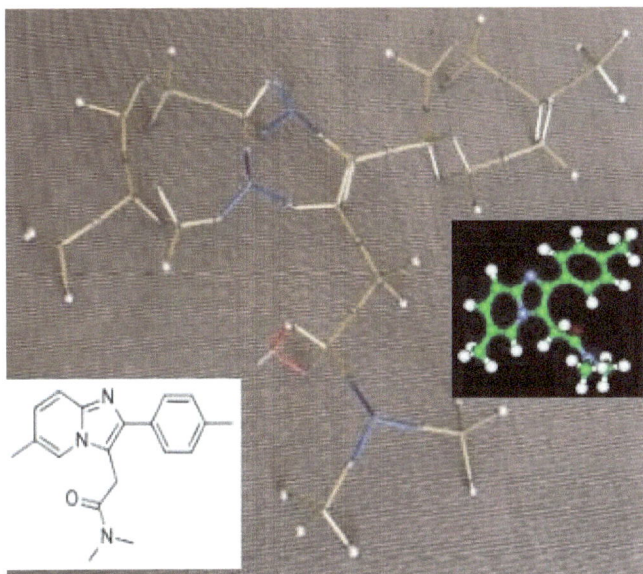

Ambien or Zolpidem is a prescription medication used for the short-term treatment of insomnia, as well as some brain disorders.

Anesthetics

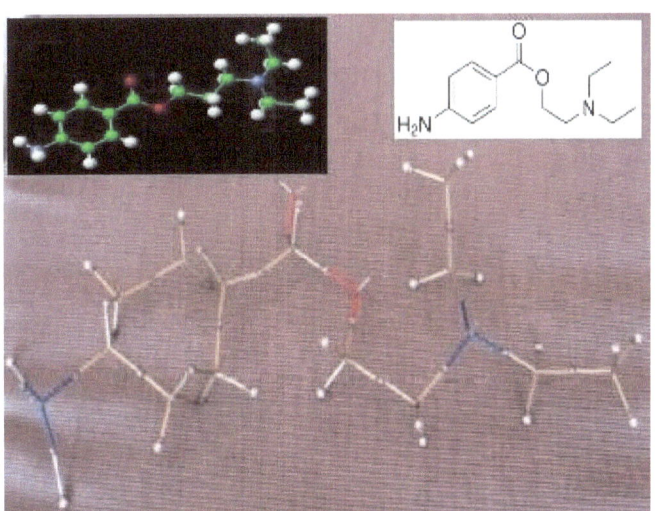

Procaine also known as Nonocaine was first synthesized in 1905, and was the first injectable man-made local anesthetic used.

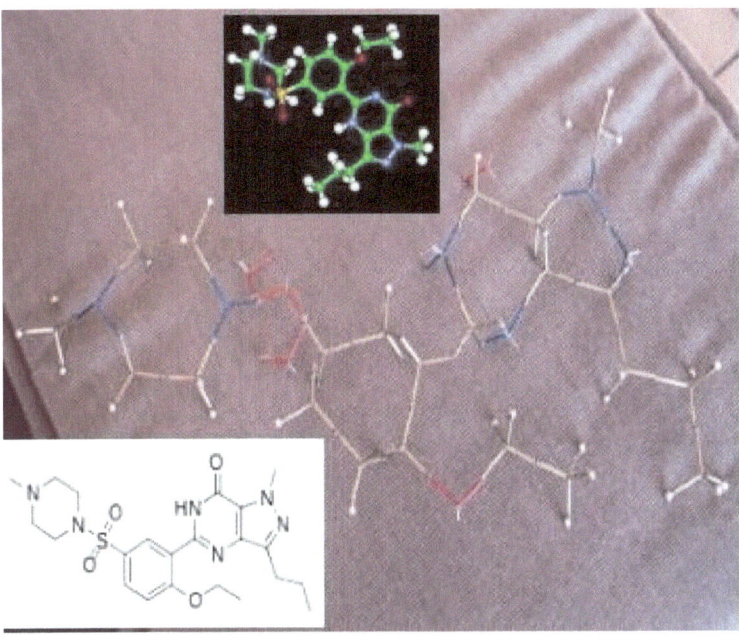

Anti-impotence agents

Sildenafil citrate, sold under the names **Viagra**, **Revatio** and under various other names, is a drug used to treat male erectile dysfunction (impotence) and pulmonary arterial hypertension.

Hallucinogens

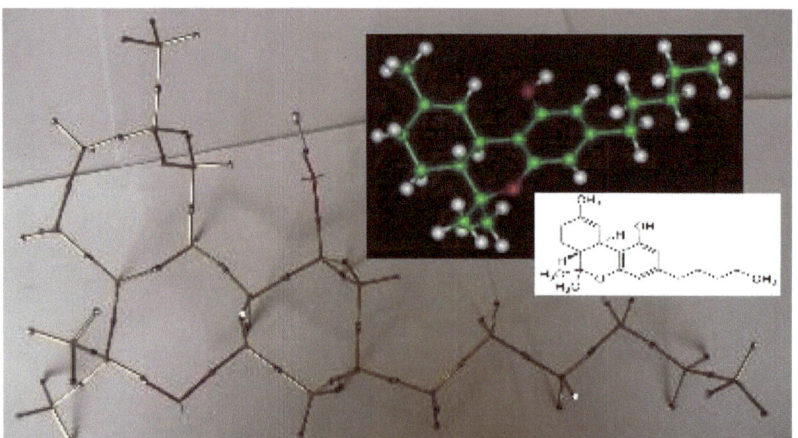

Tetrahydrocannabinol, (THC) is the main psychoactive substance found in the Cannabis plant. In pure form, it is a glassy solid when cold, and becomes viscous and sticky if warmed. As is the case with nicotine and caffeine, the role of THC in Cannabis is to protect the plant from herbivores or pathogens. THC has analgesic effects that, even at low doses, cause a high, thus leading to the fact that medical cannabis can be used to treat pain. Other effects include relaxation; euphoria; altered space-time perception; alteration of visual, auditory, and olfactory senses; disorientation; fatigue; and appetite.

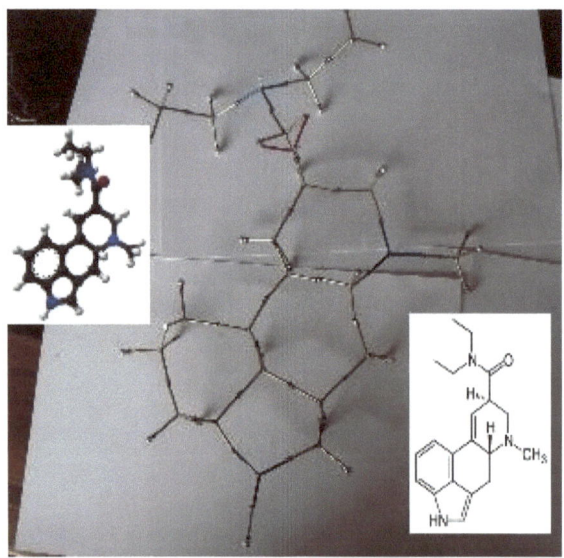

Lysergic acid diethylamide, (LSD), is probably the best known psychedelic, it has been used mainly as a recreational drug and a tool to supplement various practices for transcendence, including in meditation. It is synthesized from lysergic acid derived from ergot, a grain fungus that typically grows on rye. Some psychological effects may include an experience of radiant colors, objects and surfaces appearing to ripple or "breathe," colored patterns behind the eyes, a sense of time distorting (time seems to be stretching, repeating itself, changing speed or stopping), crawling geometric patterns overlaying walls and other objects, morphing objects, a sense that one's thoughts are spiraling into themselves, loss of a sense of identity or the ego [known as "ego death"], and powerful, and sometimes brutal, psycho-physical reactions described by users as reliving their own birth.

Crystals and Molecules in the earth

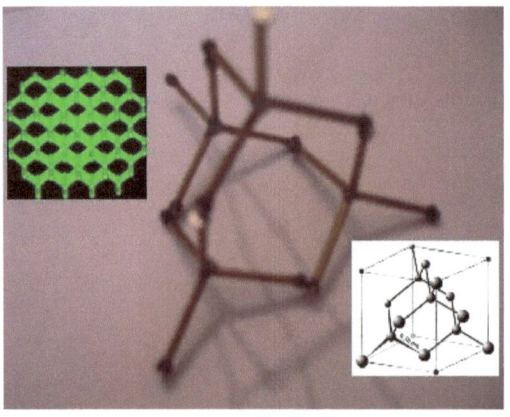

Diamond is a metastable allotrope of carbon, where the carbon atoms are arranged in a variation of the face-centered cubic crystal structure called a diamond lattice. Diamond is less stable than graphite, but the conversion rate from diamond to graphite is negligible at standard conditions. Diamond is renowned as a material with superlative physical qualities, most of which originate from the strong covalent bonding between its atoms. In particular, diamond has the highest hardness and thermal conductivity of any bulk material. Those properties determine the major industrial application of diamond in cutting and polishing tools.

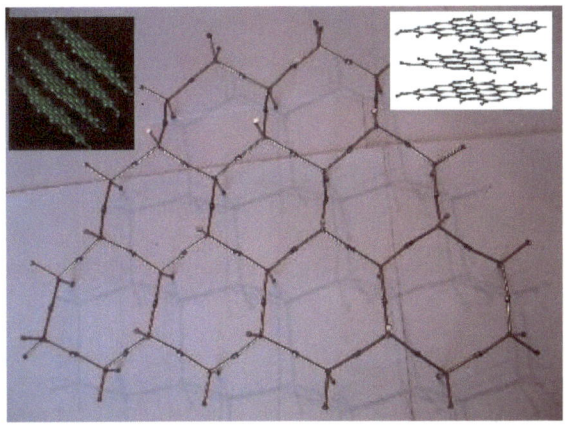

Graphite is made almost entirely of carbon atoms. It is used in pencils, where it is known as **lead.** Graphite may be considered the highest grade of coal, just above although it is not normally used as fuel because it is difficult to ignite.

Graphene is pure carbon in the form of a very thin, nearly transparent sheet, one atom thick. It is remarkably strong for its very low weight. It is 100 times stronger than steel. It conducts heat and electricity with great efficiency. While scientists had theorized about graphene for decades, it was first produced in the lab in 2004. Because it is virtually two-dimensional, it interacts oddly with light and with other materials. Graphene can be described as a one-atom thick layer of graphite. It is the basic structural element of other allotropes, including graphite, charcoal, carbon nanotubes and fullerenes. Graphene has only 3 bonds to each carbon atom. These bonds are stronger than the bonds found in alkanes and diamonds with 4 bonds to each carbon atom. This provides nanotubes with their unique strength. It is used n stealth-bomber paint because it has a super-capillary effect for trapping photons.

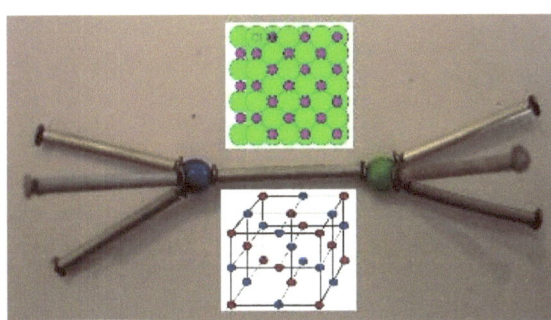

Halite is commonly known as rock salt, and is the mineral form of sodium chloride (NaCl). It is typically colorless or white, but may also be light blue, dark blue, purple, pink, red, orange, yellow or grey depending on the amount and type of impurities. With impurities removed, it is called table salt.

Molecules in the laboratory

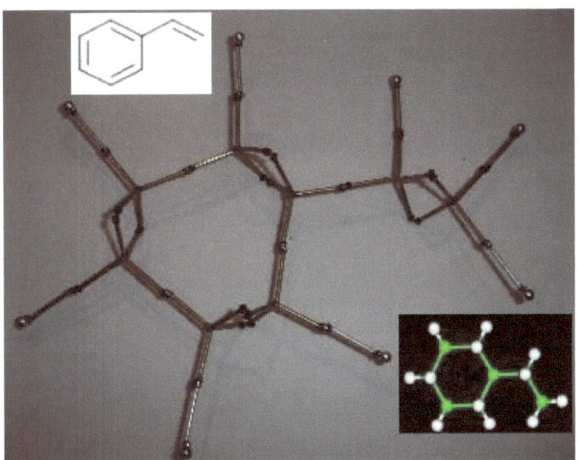

Styrene ($C_6H_5CH=CH_2$) is a derivative of benzene. It is a colorless oily liquid that evaporates easily and has a sweet smell, although high concentrations confer a less pleasant odor. Styrene is the precursor to polystyrene.

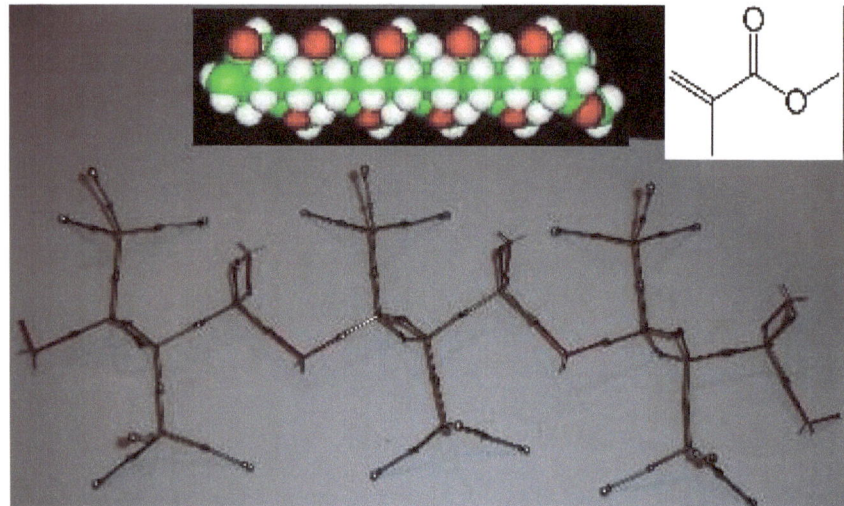

Plexiglass also called *Methylmethacrylate* $(CH_2=C(CH_3)COOCH_3)$ is a colourless liquid. It is the methyl ester of methacrylic acid.

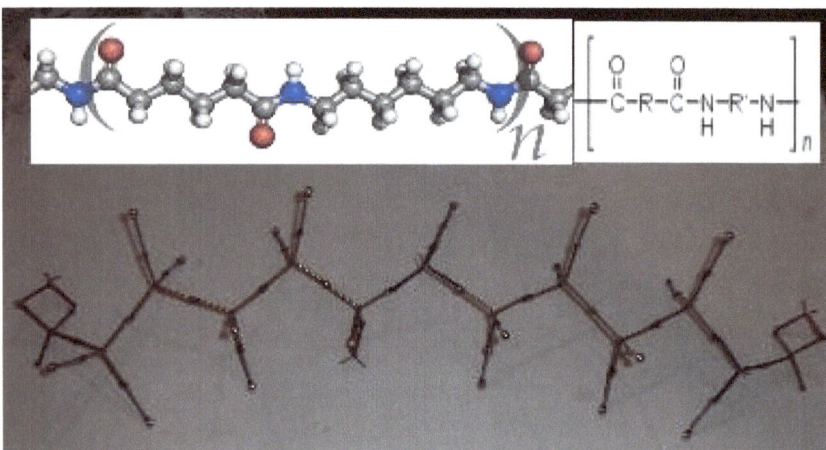

Nylon is a generic designation for a family of synthetic polymersthat were first produced in 1935.

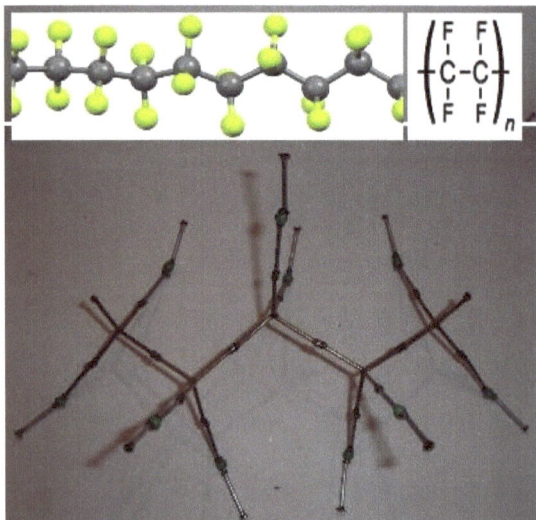

Polytetrafluoroethylene (**PTFE**) better known as Teflon is a synthetic fluoropolymer that has numerous applications. PTFE has one of the lowest coefficients of friction against any solid. It is used as a non-stick coating for pans and other cookware.

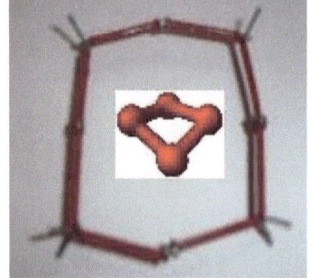

The **tetraoxygen** molecule (O_4) is also called **oxozone.** Although there are no stable O_4 molecules in liquid oxygen, O_2 molecules do tend to associate in pairs with antiparallel spins, forming transient O_4 units.

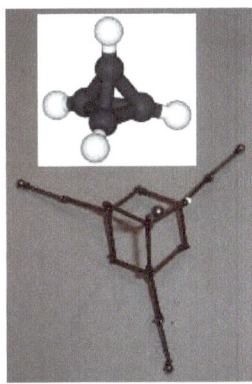

Tetrahedrane has a a chemical formula C_4H_4 and a tetrahedral structure. Extreme angle strain (carbon bond angles deviate considerably from the tetrahedral bond angle of 109.5°) prevents this molecule from forming naturally.

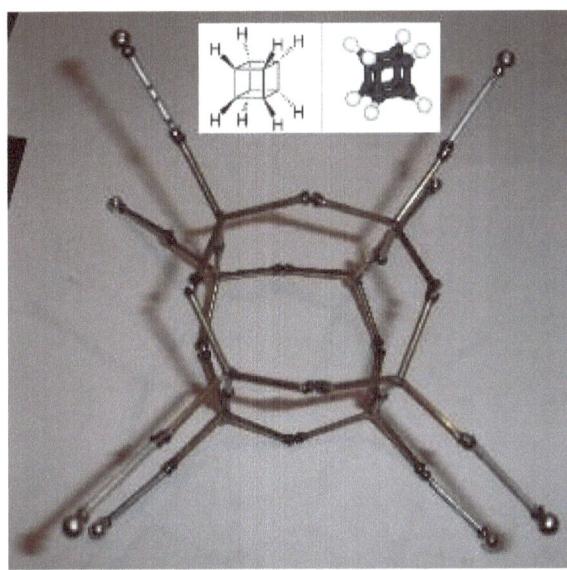

Cubane (C_8H_8) is a synthetic hydrocarbon molecule that consists of eight carbon atoms arranged at the corners of a cube, with one hydrogen atom attached to each carbon atom. It was first synthesized in 1964. It was believed that cubic carbon-based molecules could not exist, because the unusually sharp 90-degree bonding angle of the carbon atoms was expected to be too highly strained, and hence unstable. Once formed, cubane is quite kinetically stable, due to a lack of readily available decomposition paths.

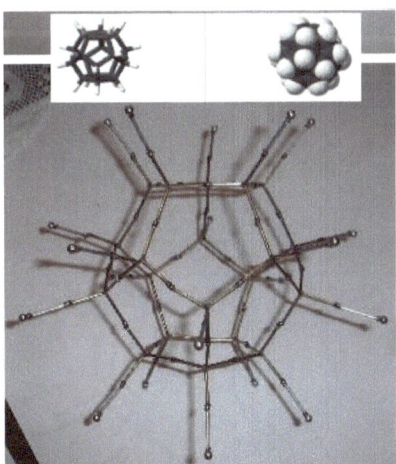

Dodecahedrane is a chemical compound ($C_{20}H_{20}$) first synthesised in 1982, primarily for the "aesthetically pleasing symmetry of the dodecahedral framework".

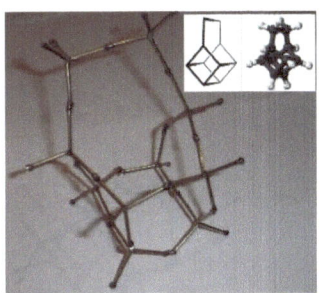

Basketane is a polycyclic alkane with the chemical formula $C_{10}H_{12}$.

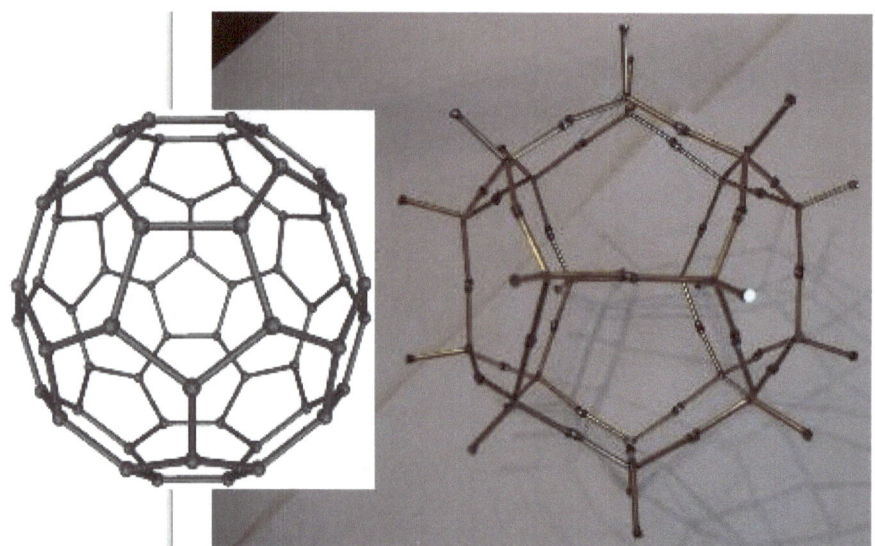

A **fullerene** is any molecule composed entirely of carbon, in the form of a hollow sphere, ellipsoid, tube, and many other shapes. Spherical fullerenes are also called **buckyballs**, and they resemble the balls used in football (soccer).

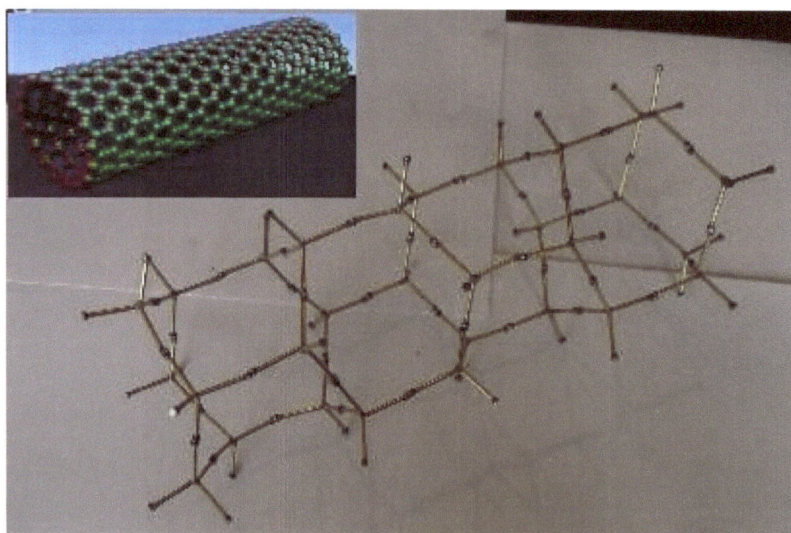

Cylindrical ones are called carbon nanotubes or buckytubes. Fullerenes are similar in structure to graphite, which is composed of stacked graphene sheets of linked hexagonal rings; but they may also contain pentagonal (or sometimes heptagonal) rings.

Comparison between various molecule modelling methods

	Ball and stick	Formula	CHONX STIX
3D	yes	no	yes
Shows double bonds	limited	yes	yes
crystals	Ball get in the way	no	yes
Ease to bild		-	easier
Ease to take apart		-	easier
Shows reactions			better
Shows protons	no		yes
Stable to hold	yes	-	no
Ease of remembering			better

There are many benefits of using CHONX STIX to learn about various molecules. With CHONX STIX, atoms are given characters with chatracteristics that represent their properties. This approach allows you to remember the properties of atoms. This makes molecules into a family of atoms bonding with each other forming bonds that are analogous to bonds people make. This approach that only CHONX STIX provides makes it easier to understand anr remember the bonds within molecules.